Illinois Foundation FFA

The Illinois Foundation FFA is the organization that provides an avenue for businesses and individuals to cooperate with the FFA in encouraging agricultural education students to improve themselves and their communities.

In 1947, a group of individuals intensely interested in the farm boys in Illinois met in Chicago for the purpose of formulating a plan, whereby Future Farmers could be given greater opportunities for the development of those characteristics essential for good citizenship and successful farming. As a result of this meeting, the Illinois Foundation FFA was created. During the years since the formation of the Illinois Foundation FFA, financial support of the program has grown from twelve companies to a base of more than seven thousand businesses and individuals.

Over seven thousand donors contribute to the Foundation with such donations totaling more than $300,000 annually. Part of this money is used to provide medals for local chapter winners and plaques for section, district, and state winners. In cooperation with the National FFA Foundation, over nine thousand chapter medals and fifteen hundred plaques are awarded annually.

Many people have described the FFA as an organization having a strong influence in building quality leadership, character, and citizenship. In looking over the rich history of the Illinois Association FFA, there is no doubt that young Americans that have evolved from the program have developed these qualities.

One of the strongest forces in the development of FFA members, has been the incentive program sponsored by the Illinois Foundation FFA. Many dedicated individuals have given their financial support. This teamwork has developed one of the finest FFA Foundations in the nation. The Illinois FFA will continue to grow as long as the cooperation between the Illinois Association FFA and the Illinois Foundation FFA remains at its present level.

Each year FFA incentive awards sponsored by the Foundation recognize outstanding accomplishment in agriculture and personal development of many members and chapters across the state. Over three thousand members participate and benefit from the incentive awards provided by the Foundation.

FFA makes a positive difference in the lives of students by developing their potential for premier leadership, personal growth, and career success through agricultural education.

Barns of Illinois

A PICTORIAL HISTORY

by Harry Porter

THE
DONNING COMPANY
PUBLISHERS

For information, please write:
The Donning Company Publishers
184 Business Park Drive, Suite 206
Virginia Beach, VA 23462

Steve Mull, General Manager
Barbara B. Buchanan, Office Manager
Jan Martin, Editorial Production Coordinator
Marshall McClure, Design Production Coordinator
Ed Williams, Project Director
Lori Kennedy, Project Research Coordinator
Richard A. Horwege, Senior Editor
Jeremy Gray, Graphic Designer
Julia Kilmer, Imaging Technician/Proofreader
Scott Rule, Director of Marketing
Travis Gallup, Marketing Coordinator

Cataloging-in-Publication Data

Porter, Harry, 1935–
Barns of Illinois : a pictorial history / by Harry Porter.
p. cm.
Includes index.
ISBN 1-57864-205-1 (alk. paper)
1. Barns—Illinois. I. Title.

NA8230.P67 2003
728'.922'09773—dc21

2003040971

Printed in the United States of America

CONTENTS

FOREWORD

Old barns are where kids worked, played, laughed, cried, thought, and learned about life. You have heard the saying, "Everything I needed to know I learned in kindergarten"? Most farm kids learned all they needed to know about life in and behind the family barn!

The barn was always a safe haven from the cold winds of winter, the hot days of summer, and a hiding place from those household chores your mother always seemed to have. Remember those days when you were rowdy in the house and your mother would tell you to play outside and how you always ended up playing in the barn or following your dad around?

Remember at an early age how being around animals you soon figured out the birds and the bees thing? What we witnessed firsthand on the farm made reproduction a natural occurrence and not a big deal. Nowadays a big fuss is often made about providing the same information in sex education classes.

Remember the old bulls and how you knew at an early age to be careful so one didn't hurt you? When the bull got agitated and mean he would be put in an extra strong box stall in the barn?

Do you remember the good times kids had growing up on the farm? Remember shooting baskets at a hoop nailed to the side of the barn? Remember walking in fields of clover and alfalfa and the sweet smell when it was cut? Remember the wonderful times you had swinging on hay ropes and building hay tunnels and playing in them for hours?

Remember the invention of the hay track, later the hay baler, and how they made putting up hay a little easier, but not much easier? Remember baling alfalfa on a hot August day and stacking bales in a hot dusty hayloft with the temperature hovering over a hundred degrees?

Remember driving a team of horses attached to the big hay rope, which was used to pull the hay up to the hay track and across the barn, and how you strained to hear those in the haymow yell whoa, so you could stop the team? However the team usually stopped before you heard the sound. Then you would back the team up for the next load, repeating the process until the harvest was in? The old farm truck later replaced the team of horses and this task became a small child's first driving experience.

Hay was at first stored loose, next came rectangular bales, followed by round bales wrapped in plastic and stored outside, making the haymow obsolete. Remember sitting in the haymow door dangling your feet outside, dreaming those big dreams that all kids have, watching the animals, and seeing all those exciting things that the others were doing and hoping nothing would ever change but knowing it would?

Have you ever experienced solitude while watching a cow chew her cud as a cold rain came down outside or had the joy of finding a new litter of kittens in the hayloft guarded by their ever alert mother? Do you recall hearing the flutter of pigeon wings in the cupola when some strange noise, usually a false alarm, startled them into flight, or the thrill of seeing an owl, sitting on a rafter waiting patiently for it's next meal, usually an unsuspecting field mouse?

Remember the cured hams that hung in the rafters of the oats bin in the barn, hanging there until they were eaten?

Remember all the assortment of pets that was on every farm that we all dearly loved. Remember all those mornings when you walked out of the house to start the morning farm chores, being led to the barn by the ever-faithful family dog, full of cockleburs and matted hair, with its constant wagging tail, thinking perhaps it was his duty to make sure you had safe passage to the barn?

Do you remember going outside in the dark of an early winter morning as the mooing cows waiting to be fed and milked, directed you to the barn?

Do you remember squirting milk directly from the cow at one of your siblings or into your own mouth? Remember filling a pan with milk for the cats that lived in the barn and helped rid it of rodents?

Do you remember the smell of animals, the old harness and saddles, fresh milk, manure, hay, and straw all blended together? Smells we will long remember and hopefully never forget.

Remember sneezing while spreading straw in a box stall for the expected arrival of a new calf?

Do you remember cows standing in opposite directions, swatting flies from each other's heads with their tails moving in rhythmic unison, almost as if music was playing that only they could hear?

Remember finding a spider web where none had been the day before and watching the spider expanding its web to catch those pesky flies?

Remember the barn swallows that lived in the rafters of the barn, and how you learned not to shoot them with your Daisy B.B. gun, because Dad said they were good birds and only ate insects?

Do you remember the names of the draft horses, the family dog, and the cows that were milked everyday, morning and evening?

Hopefully this book will help bring back fond memories of a wonderful time in your life back on the farm and help you recall all the good times you had in the barn of your childhood.

PREFACE

Farmers, as well as the Illinois FFA, recognize the contributions barns have made in the past, but also recognize the changes in farming practices that have made many barns obsolete. An owner wanting to preserve a barn often does so only at great expense. Barns were built to last for generations and last they did. Some were built before the Civil War and many of those are still in use today.

Barns once central to farm life now have empty horse stalls. The milking parlor is no longer used and the stanchions serve only as a support for spider webs. The smell of animals that once engulfed the barn often is just a lingering memory. The old weathered barn, though empty, still has character and resilience. If you listen carefully, you may still hear the echoes of contented sounds of animals that lived here long ago mingled with the laughter of young children at play.

Many barns now stand empty. Holes in their roofs have allowed the rains to come in, causing the rafters to sag. Flocks of birds fly in and out unimpeded. The little hay that remains in the haylofts is wet and decaying. While some barns stand straight and freshly painted and are still being used for the purpose for which they were built, some barns have weatherboards missing, leaving holes that expose more of the interior to the elements. These barns still hold just as much interest to a true barn lover.

If you revisit the old barn that you knew as a child, though it may have changed and seem smaller, remember your childhood experience, and hopefully you will find it as a delightful a place as it was back then.

The old Illinois barn's crucial role in farming may have come to an end, but some barns now enjoy a new usefulness as retail stores, storage sheds, museums, churches, homes, restaurants, theaters, and resorts. However, over the years, alternative use of barns has not been very successful. The expenses of maintenance, repairs, and real estate taxes have hastened their demise.

The farm gave us our livelihood, roots, history, and family values. The family barn, so central to farm life, played a big part in developing those values that we all hold so dear and hope we never lose. Barns project a sense of history and integrity, and remind us of a time when a handshake was binding.

Nothing lasts forever—not even barns—though these great structures have withstood the elements and the test of time for generations. An increased interest in preservation must be developed or we will lose most of the great barns of Illinois. It is hoped that in some small way that books like *Barns of Illinois* will help further the interest in preserving these great edifices of rural Illinois.

The *BARN AGAIN PROGRAM*, sponsored by *Successful Farming Magazine* and the National Trust For Historic Preservation, has been largely responsible for the increase in barn preservation.

AUTHOR'S ACKNOWLEDGEMENTS

This book, *Barns of Illinois*, resulted from a request by the Illinois Foundation FFA for barn pictures and stories about barns. This book is not a tourist guide, or a road map to the great barns of Illinois because too many great barns are not featured here.

Past and present FFA members and others submitted the material used. FFA members in conjunction with the Illinois Foundation FFA have worked diligently providing pictures and captions. This activity provided an opportunity to learn about a time when the barn was the center of farm life.

The public response to this project has been overwhelming and beyond expectations, so many barn pictures and stories have been submitted that not all could be used. The plans are to retain the submitted pictures and material, not used in this volume, and utilize them in proposed future volumes.

In publishing this book we have relied primarily upon those who submitted the pictures for information such as location and description of the barn. Some words have been used interchangeably such as *haymow* and *hayloft*.

This book is dedicated to those that are true barn lovers, especially those who contributed photographs, stories, and information about each featured barn. The contributors who generously shared information and photographs are greatly appreciated. With out their help this book would not have become a reality.

As author, I would like to thank some very special people who helped me make this book a reality: My wife Carol, my two sons Steven and Scott, my daughter-in-law Kim, my faithful sister Wilma Gay, and Frank Dry, associate executive director of the Illinois Foundation FFA. A deep sense of gratitude goes out to Donald H. Scott, Ken Kashian, Bill Scott, Lois Herr, Paul Phelan, and Kent Slater for their help support and encouragement.

FFA ACKNOWLEDGEMENTS

The FFA members would like to thank all those people that permitted them to go upon their land and photograph the great barns of Illinois and provided them with the information in the captions.

The Illinois Foundation FFA would like to express thanks to their staff for all the work they did in making this book a success. Without their help this book would never have been completed.

We also want to express our appreciation to Donning Company Publishers for their professional guidance. They assisted us in an endeavor in which none of us had any previous experience.

The Illinois Foundation FFA would like to thank all those people that submitted photographs that are not featured here. It is our hope that they may appear in proposed future volumes. To Peg Kaye a heartfelt thanks for helping us receive pictures and information.

The Illinois Foundation FFA would also like to thank the Association of Illinois Electric Cooperatives. A special thanks goes to John Lowrey, editor of *Illinois Country Living Magazine*, and to Farm Credit Services for their financial support.

Illinois Foundation FFA Building.

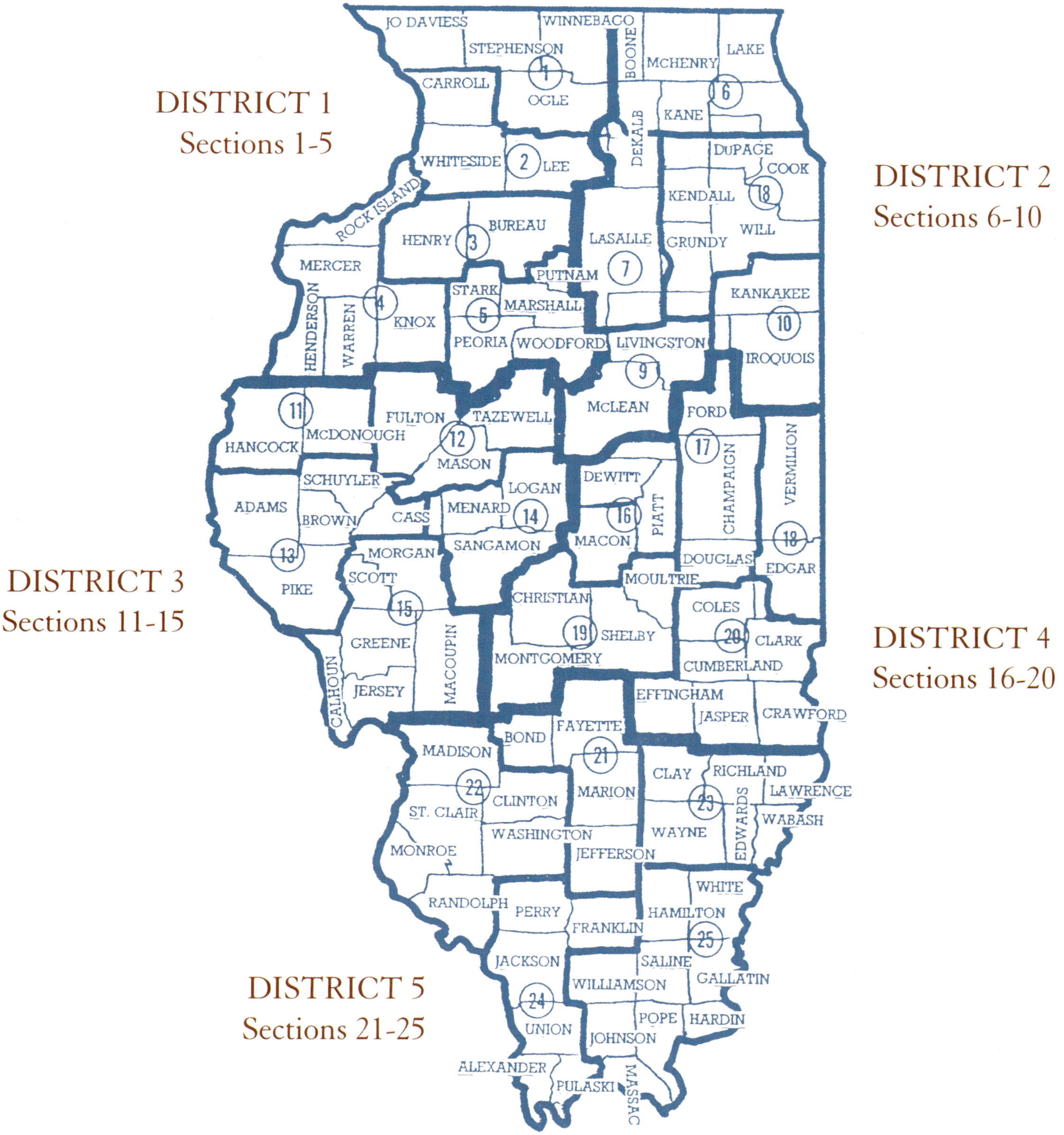
DISTRICT 1
Sections 1-5
DISTRICT 2
Sections 6-10
DISTRICT 3
Sections 11-15
DISTRICT 4
Sections 16-20
DISTRICT 5
Sections 21-25
JO DAVIESS
WINNEBAGO
STEPHENSON
BOONE
MCHENRY
LAKE
CARROLL
OGLE
DEKALB
KANE
WHITESIDE
LEE
DUPAGE
COOK
ROCK ISLAND
KENDALL
HENRY
BUREAU
LASALLE
GRUNDY
WILL
MERCER
PUTNAM
KANKAKEE
HENDERSON
WARREN
KNOX
STARK
MARSHALL
PEORIA
WOODFORD
LIVINGSTON
IROQUOIS
HANCOCK
MCDONOUGH
FULTON
TAZEWELL
MCLEAN
FORD
MASON
VERMILION
CHAMPAIGN
SCHUYLER
DEWITT
ADAMS
BROWN
CASS
MENARD
LOGAN
PIATT
MACON
MORGAN
SANGAMON
DOUGLAS
EDGAR
PIKE
SCOTT
MOULTRIE
CHRISTIAN
COLES
GREENE
SHELBY
CLARK
MACOUPIN
CALHOUN
MONTGOMERY
CUMBERLAND
JERSEY
EFFINGHAM
JASPER
CRAWFORD
MADISON
BOND
FAYETTE
CLAY
RICHLAND
LAWRENCE
MARION
CLINTON
ST. CLAIR
WASHINGTON
WAYNE
EDWARDS
WABASH
MONROE
JEFFERSON
WHITE
RANDOLPH
PERRY
FRANKLIN
HAMILTON
JACKSON
SALINE
GALLATIN
WILLIAMSON
UNION
POPE
HARDIN
JOHNSON
ALEXANDER
PULASKI
MASSAC
1
2
3
4
5
6
7
8
9
10
11
12
13
14
15
16
17
18
19
20
21
22
23
24
25

District 1, Stephenson County

This red barn with a white fence was built in the 1860s. It represents the German Forebay/Bank barn design. The barn is unique in that its dimensions are almost square. This barn is located in Harlem Township west of Freeport.
Photograph and information submitted by Judith Gonder.

District 1, Whiteside County

This large barn, called the Nermes-Penrose Barn, was built around 1900. It sits on farmland once donated to KSB Hospital by the hospital's founder, Judge Bethea. Ed Hermes is the current owner of this barn located northeast of Sterling.
Photograph and information submitted by Roselyn Nesbit.

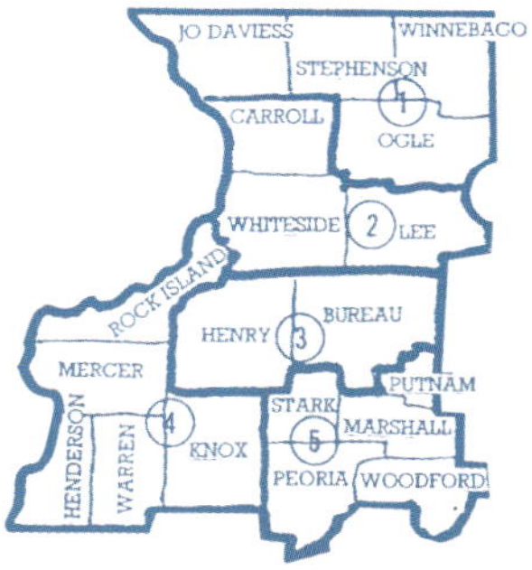

DISTRICT 1

District 1 is in the northwestern part of Illinois and was the last portion of the state of Illinois to be settled. The majority of the settlers came from the eastern part of the United States from states like Ohio, Pennsylvania, New York, etc. The majority originally were immigrants from countries in Northern Europe such as Belgium, Norway, and Sweden, etc.

The Mississippi River, which borders the western part of the territory, has had a tremendous influence on the region. Early on the influence was in the formation of the types of soil. Today, it provides a means to transport many tons of corn and soybeans produced in the region, as well as being the vehicle of transportation for other commerce. An example of this would be the Quad Cities once known as the machinery capital of the world.

The gently rolling to level farmland lends itself to intensive livestock production—dairy in the northern part, swine concentrated in the central part, and beef cattle in the southern area. Livestock production necessitated more forage crops and small grains in the land use program, which was used for feed and pasture. There was a greater need for housing to protect the livestock from the cold winters, thus the variety and numerous styles of structures such as barns and sheds for windbreaks.

Farms became larger and there became fewer farmers. At the same time manufacturing of farm machinery began to diminish. Tourism has now become the number one income-producing entity of that part of Illinois. Agriculture has slipped to number two. However, preserving past agricultural methods tends to give credence to the increasing number of tourists.

—Eldon R. Aupperle
Toulon, Illinois

District 1, Whiteside County

This barn, located on the outskirts of Coleta, has served past owners well, although no repairs have been made for years. The old barn ravaged by time and elements is shown here standing square and plumb—a credit to its builder.

Photograph submitted by Melissa Gascoigne.

District 1, Henry County

Myron McHenry purchased this farm in 1890 and trained trotting horses here until 1906 when the Hennepin Canal went through the property. One of the most famous trotting horses to train here was "Dan Patch." Dean and Diane Olson and family are the current farm owners. There were other barns on the property similar to this one that served as living quarters and home for the jockeys. One side of the barn floor was dirt, the other side wood, with a concrete floor down the middle. There originally were twelve box stalls in this barn—one remains intact. Theodore Edman owned the farm prior to Carl Olsen who purchased it in 1963.

The detached summer kitchen was also built in the 1800s. There was no kitchen in the home. The hired help lived on the second floor of this building. The barn and the detached summer kitchen are located on Grange Road north of Geneseo.
Photograph and information submitted by Drew De Reiemacker.

District 1, Jo Daviess County

This old barn sits on a slope overlooking a pond in one of the many beautiful valleys in Jo Daviess County, just below the lookout tower, a popular tourist stop on Route 20, near Galena, Illinois.

Photograph and information submitted by Jeff Slamp.

District 1, Lee County

These two barns of two different styles were photographed in evening light. They have been maintained in exceptionally good condition on this working farm. These barns, if maintained, should last a hundred years or more. The barns are located south of Dixon.

Photograph and information submitted by Phil Mattox.

District 1, Henry County

This barn, sitting on a rise along a country road just off of I-80 near Annauan, is still showing pride with the American Flag blowing in the breeze. The picture says more then ten thousand words ever could. This barn is a reminder of the past and the traditions it was built on. Note the shadow of the companion windmill on the barn in the top photograph.

Top photograph submitted by Paul Phelan, lower photograph submitted by Courtney Reiling.

District 1, Woodford County

This green-roofed barn with its red brick foundation seems to be smiling, even after working hard for so many years for its owners. Notice the wooden ramp. Usually a ramp would appear on both ends so a team and wagon could go straight through the barn and out the other side. This barn is located near Washburn.
Photograph submitted by Joe Steffen.

District 1, Jo Daviess County

This barn was located on Greenvale Road near Greenvale. It sat on a scenic hillside, which overlooked the Apple River Valley. Recently, this barn blew down.
Photograph submitted by Randy Marks.

District 1, Whiteside County

Henry R. Mammen built this barn in the 1920s. The amazing thing is that he built this barn with just one arm, having lost the other one to disease. This barn is located just west of Sterling.
Photograph and information submitted by Frances Schott.

District 1, Whiteside County

Henry R. Mammen is the owner of this farm. It is the birthplace of Frances Schott. This barn, located near Agnew, was built in 1918.

Photograph and information submitted by Frances Schott.

District 1, Winnebago County

This January 2002 photograph shows the south side of the Weldon farm after a frosty night and an early morning snowfall. The Weldon farm, located seven miles west of Rockford, has been owned by one family since it was founded in 1849. The principal barn and attached carriage house was built about 1885. Until the 1920s the farm was a Percheron and French coach horse breeding farm. In 1944 the barn was converted to a dairy barn with thirty stanchions. Randy and Pat Bruer manage the dairy.

Photograph and information submitted by Bob Weldon.

District 1, Lee County

Henry W. Hew built this great white barn in 1917. It was the home of the Hill Den Dairy. The barn is located on Hill Den Farm on Palmayra Road near Dixon.

Photograph submitted by Helen Rodgers.

This is a great companion barn. Fred Lard built this barn in 1857 on the Hill Den Farm on Palmayra Road near Dixon.

Photograph submitted by Helen Rodgers.

District 1, Whiteside County

The Ed Hermes family has owned this barn northeast of Sterling for decades. Someone sought to purchase it for a home. It never became a reality because it was destroyed by a recent fire of undetermined origin.

Photograph and information submitted by Roselyn Nesbit.

District 1, Ogle County

This well-maintained red barn and companion buildings are located near Lynnville. The turn into the drive makes for a great fall scene.

Photograph submitted by Ken Kashian.

District 1, Stark County

This red barn with its white trim was built circa 1930. It measures thirty-four by forty-two feet. It has eight horse stalls, a box stall, and grain storage bins. One side of the barn was used for milking cows and housing calves. The other side was used for working horses and later riding horses. A unique feature of this barn is the wooden stairs leading to the hayloft. The barn is located 3.5 miles southwest of Bradford.

Photograph and information submitted by Karen E. Andrews.

District 1, Mercer County

The Willits family are the proud owners of this Carpenter Gothic design barn on their farm called "Verduree." The barn has an unusual curved top swinging driveway door and two tall side doors as well as Gothic trim. This barn is located between Keithsburg and New Boston.

Photograph and information submitted by Kent Slater.

District 1, Whiteside County

Clarence Wolf built this white barn in the 1940s for its former owner Edward Schott. Mr. Schott also helped in the construction. The farm was sold to his son Laurence, now deceased. The present owner is Frances Schott. This barn is located three miles north of Prairiesville.

Photograph and information submitted by Frances Schott.

District 1, Winnebago County

Kathie Jurowski poses with a heifer calf in front of her barn. She and her husband, Jerry, raise unique Belted Gallaway cattle as a hobby. The barn is a classic with milk-cooling house attached. But the hardy Scottish-bred Gallaways prefer to stay outdoors. This barn is located on Yalebride Road in Rockton.

Photograph and information submitted by Gail S. Ravitts.

District 1, Ogle County

The Rall brothers of Kings, Illinois, built this round-roofed barn in 1920. This is one of the first barns to be built with this type of round roof. The barn was originally built with horse stalls on one side. The family milk cows were on the other side. The barn is still used to store hay and straw and as a pigeon haven and a refuge for the family dogs. The current owners, John and Norma Swanson, purchased this farm in 1968. The barn is located on Grant Road near Chana.

Photograph and information submitted by John and Norma Swanson.

District 1, Stephenson County

This is a barn that smiles on a daily basis. The barn, located north of Pecatonica, Illinois, is owned by the Carroll family.
Photograph and information submitted by Sharon Hull.

District 1, Stephenson County

The barn is striking not only for its blue color but also for the setting. The barn's location south of Freeport provides a panoramic view of many farms on gently rolling terrain. It is picturesque in any season.
Photograph and information submitted by Roselyn Nesbit.

District 1, Bureau County

This barn is located southeast of Walnut. It was used for milking cows and later for cattle and hog production.
Photograph and information submitted by Doug Anderson.

District 1, Jo Daviess County

The late William Upman was the owner of this great old barn with its shinning aluminum roof located at the bottom of a gradual slope called Hazy Hollow. Mr. Upman used this barn as a milk barn for many years. Now the barn is only used for storage of hay. This barn is located in upper Jo Daviess County.
Photograph and information submitted by Randy Marks.

District 1, Winnebago County

This former barn is located at 411 Kent Street in Rockford, Illinois, at the Tinker Swiss Cottage Museum. In 1998 the reconstruction of the historic Tinker barn began. Using well-seasoned beams from old buildings in the Rockford area and elsewhere, the structure was formed in the timber frame manner, using wooden pegs as fasteners. Siding from barns in Illinois and Wisconsin was used on the interior walls of the upper floors. Basement walls were covered with a limestone veneer and rustic wood. When completed, the barn will contain a visitor center and museum store, curator's workroom, and an all-purpose room suitable for parties, children's activities, and receptions.

Photograph and information submitted by Tinker Swiss Cottage Museum.

District 1, Marshall County

This red barn is sporting a new paint job. Located at Sparland, this barn has been in the same family since 1919 and is used to raise Black Angus cattle.

Photograph and information submitted by Jim Frank.

District 1, Ogle County

The photographer says that this was the first barn picture she took when she first got interested in barn photography and has been hooked ever since. She liked the contrast of the horizontal foundation boards with the upper vertical boards, and the ramp leading up to the barn. This barn is located on the Stillman Valley Road east of the Rock River.

Photograph submitted by Joan Eshleman Velde.

District 1, Ogle County

This old barn sometimes gets overlooked when people pass by because it is behind another old barn. It looks a little sad with its hayloft door hanging by just one hinge. This barn is located on the Stillman Valley Road east of the Rock River.

Photograph submitted by Joan Eshleman Velde.

District 1, Henry County

This beautiful late evening winter scene shows a red barn and glazed block silo. This barn, located two miles north of Lena, was built in 1900 with a silo. The overhead door shed on the right was added in 1976. The lean-to, on the left, was added in 2001. The barn has served as a cow and calf barn, and as hay and lumber storage. It also has a shop and horse-drawn vehicle storage area, and serves as a horse barn. Photograph and information submitted by Wieland J. Kayser.

District 2, Will County

This barn was built in 1950 by Clarence G. Rowley, his son Elmer W. Rowley, and his grandson Edward R. Otto. The original barn did not have a hayloft, so ten years later it was added. This farm, located near Lockport, has been in the Rowley family for over a hundred years.

Photograph and information submitted by Joanne R. Harvey.

District 2, Kane County

This barn, located on Harmony Road near Hamshire, is owned and used by the Stoxen family.

Photograph submitted by Alissa Stoxen, information submitted by Donald Stoxen.

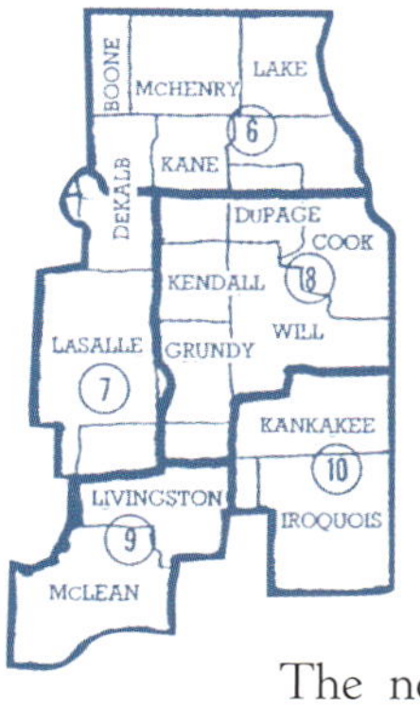

DISTRICT 2

The northeastern corner of Illinois is home to District 2 of agricultural education. The area is predominantly known for the Chicago metropolitan area, but still has a viable farming base that is some of the most productive in the entire country.

The original skyscrapers of this region were the barns that allowed farming and the livestock industry to flourish. The magnificent structures may be shorter in structure than the modern-day office buildings, but are no less marvels of architecture.

Although grain farmers are now dominating the farming of the region, the remaining barns of northeastern Illinois are testament to a once thriving livestock industry of the area. Dairy barns, especially in the district's northern border, dominate the rural landscape that makes many a traveler take a second look and to take a little extra time to enjoy these great structures.

The barn architecture of the area reflects many different cultures. Dutch, German, and Norwegian settlers were pioneers of farming in the area. In fact, the first permanent Norwegian settlement in America is located in LaSalle County. Our Dutch ancestors are responsible for barns that seem to have the "tips" of the roofs cut off.

—Jeff Maierhofer
Agricultural Instructor, Seneca High School

District 2, Livingston County

For decades this seemingly abandoned old barn, with its partial rusting metal siding, has been a landmark for travelers. This barn is located south of Dwight.

Photograph submitted by Tom Madigan.

District 2, La Salle County

This barn, located near Ransom, is owned by Paul A. Phelan. Paul, an avid barn photographer, contributed numerous pictures to be considered for this book. This barn was built in 1901 with peg and beam construction, tongue and groove wood siding, and a cedar shingle roof. The barn is sixty feet long across the front face and has three fourteen-foot bays making it forty-two feet wide. It has a feed crib with a capacity of twelve hundred to fifteen hundred bushels of corn across the west end of the barn. Ten feet out from the barn wall is an eighteen-by-sixty-foot lean-to. The crib wasn't as long as the barn was so there was a small pen at either end of the crib. The north pen contained a Letz burr mill and there was a small door in the west wall to run the belt through to drive the grinder. The crib fell victim to a windstorm in the early 1960s and was replaced with an eighteen-by-sixty-foot lean-to with salvaged material from the crib and used telephone poles and metal siding and roof. I remember one team of horses in the barn but there was room for six horses in the south side of the barn with a feeding alley in front of them. There was a feed storage bin across the center of the east face of the barn. There was another feed alley on the north side of the barn that ran in front of a group of six cow stanchions that were homemade. A few cows were milked twice a day for home use and cream production for grocery barter. There was a box stall on the north side for feeding cattle. A few pigs were kept in the alleyway between the barn and the feed crib and in a lot on the south side of the barn and feed crib and in a lot on the south side of the barn. The barn was re-roofed with cedar in the 1930s and then with metal in the mid-1980s. The wood siding was covered with metal in the mid-1980s. The wood siding was covered with corrugated galvanized metal after World War II. There was some foundation repair on the west wall and siding repair in the mid-1950s. The horse stalls were removed and the floors cemented and used for hog production through the early 1980s, then for sheep production until 1992. It is now used for small machinery storage and tile fitting storage for my son John's drainage business.

Photograph and information submitted by Paul Phelan.

District 2, Kankakee County

The James Manufacturing Company of Wisconsin constructed this barn in 1929. It was originally built as a dairy barn for Julia Cornstock. At that time it was one of the largest wooden barns in Illinois. Ruthie and Darryl Kleinest purchased the farm in 1978. On February 5, 1979, the barn and over one hundred head of sheep were destroyed by fire. The barn was replaced (see photographs below) by enlarging and rebuilding a nearby horse barn. The new barn is located near Chebanse.

Photograph and information submitted by Brenda Wasilewski.

The farmstead with the original barn before the fire.

Repairing and adding on to the damaged barn after the fire.

A close-up look at the new, completed barn.

The finished barn from a distance.

District 2, Grundy County

The only information available is that this barn, located near Channahan, is called the "Dollinger Mule Barn."

Photograph submitted by Ken Kashian without information.

District 2, Will County

This barn and the rainbow were photographed on a fall day with the rain clouds in the background. Unfortunately, the barn was demolished shortly after this picture was taken. This barn was located near Wilmington.

Photograph and information submitted by Jo Anne McIntire.

District 2, Boone County

A fall storm was brewing when this photograph was taken. Ottello Domenella is the owner of this barn. The barn, located three miles north of Harvard, was on the farm when purchased. Photograph submitted by Mildred Banks.

District 2, La Salle County

A portion of this barn was once a Lutheran Church in the town of Stavanger. It was moved to its present location and converted into a barn. The right side of the barn was already a barn which was moved and then attached to the church portion making a great barn. This barn is located near Stavanger.

Photograph and information submitted by Paul A. Phelan.

District 2, McLean County

This brightly colored barn seems to be smiling for the camera. The farm is known as the Wiebel Farm. The companion silo helps make for a great picture of an Illinois barn. This barn is located south of Cropsey.

Photograph and information submitted by Paul Phelan.

District 2, Lake County

This barn is known as the Brogden Family Barn. The barn was built in 1864. The silo and milk-cooling house were built around 1900. This barn is located in Liberty Township.

Photograph and information submitted by Nancy Burgess.

District 2, La Salle County

This unusual old barn, located south of Ransom, is still being used today. The barn is known as the Corrigan Barn and is also referred to as the Frank Fransico Barn, after the current user of this barn. It is an imposing site coming from the east and is often photographed. It is sixty feet long, forty feet wide, and at least twenty-four feet to the eves. The barn has board and batten siding and an asbestos roof. The date of construction is not definitely known but a sketch of the house and barn appeared in a County Atlas or Plat Book in 1873. It has three grain bins along the north wall then an alleyway. There are two areas in the rest of the barn where draft team stalls were located. The feed, tack, and supply rooms are still there but the stalls have been removed. There is a driveway between these two areas where loads of hay were driven through and conveyed into the upper storage area. To stand in this alleyway now and look up into the empty barn with its distinctive configuration with the two cupolas is like looking up into a cathedral.

Photograph and information submitted by Paul Phelan.

District 2, McHenry County

This hip-roof barn, located six miles northwest of Marengo, was built in 1910. The barn has a cement silo inside and a companion silo outside. McKee Bros. & Sons is painted above the barn's main door.

Photograph and information submitted by Chris McKee.

District 2, Boone County

Mildred Banks photographed this red barn with its white trim as she drove through the country on an early July evening. The barn, owned by Kevin Stark, is located eight miles from Belvidere. The photograph and information submitted by Mildred Banks.

District 2, Kendall County

This barn was built in 1979 and used as a dairy barn. It holds forty-two cows and measures 36 by 105 feet. In 1994, the haymow, which was half full with hay collapsed in the area where the cows were kept. Fortunately, none of the cows where in the barn at the time and there were no injuries. The cows were transported to a neighboring dairy farm five miles away where the milking was done twice a day until repairs could be made. The barn was back in operation within four months. A 4,000-gallon milk-holding tank was added in 2002. Kirk Friestad is the owner of this barn located three miles north of Central at the junction of Route 47 and Route 52.

Photograph and information submitted by Joe Steffen.

District 2, McLean County

This barn was built in 1890. It housed horses that did the fieldwork. Later, Mr. Brady the owner who was a Funks Seed Corn dealer used it to store seed corn. It now provides shelter for the swine that are part of 4-H and FFA projects. The barn is located on the George Brady Centennial farm near Chenoa.

Photograph and information submitted by Sheryl Rosenberger.

District 2, Will County

This typical barn on the Illinois prairie is special to its owner. The barn, built before 1900 with its companion silo, was a dairy barn from 1940 through 1955. Since then it has been used as a storage building for farm machinery. The barn is located near Lockport.

Photograph and information submitted by Evelyn Kolar.

District 2, Will County

Melvin and Esther Heisner built this barn near Peotone in 1962. It is unique because of the large shingle artwork on its roof. The farm name outlined in shingles on the roof has been a landmark for travelers and local residents. Truckers and airline pilots have used the Mel-Est barn as a reference point. The barn has a sixty-cow capacity and houses the Mel-Est registered Holstein herd. Some of the registered Holsteins that were raised here grew up to be known around the world.

Photograph and information submitted by Melvin and Esther Heisner and their five sons.

District 2, Livingston County

This horse barn was built between 1910 and 1920. It has ten horse stalls with room for wagons and carriages. Later on, one end was converted for machine storage. The barn has a three-inch plank floor approximately two feet off the ground. This barn is located about five miles north of Colfax, near Fairbury.

Photograph and information submitted by the Henry Winter Land Trust, Attention: Brad Brown.

District 2, Boone County

This barn was used as a dairy barn until the 1950s when Bob and Betty Edwards switched their focus to apple orchards. In 1964 they transformed the barn into the "Apple Barn." They sold apples, pumpkins, apple pies, and apple donuts. Going to the Edwards' apple orchard became a Rockford tradition. The Edwards' daughter Barb and Ken Hall continue to fill the "Apple Barn" with the aroma of apples and cinnamon. The barn is located near Popular Grove.

Photograph and information submitted by Gail S. Ravitts.

District 2, Boone County

This red barn with its green roof and attached milk-cooling house is now part of the "Pumpkin Patch," a miniature theme park open from October through December. A huge selection of pumpkins along with pony and wagon rides and creative ideas are offered. The barn is located near Caledonia.

Photograph and information submitted by Gail S. Ravitts.

District 2, Will County

I am happy to tell you about our barn because I hold the FFA American Farm Degree, which I received in 1952. I hope you enjoy reading about our barn's unique history. The late Senator Richard J. Barr, who was an Illinois senator for forty-eight years, left the farm as a young man, but always kept a keen interest in it. His pride was the good cattle which were fed on the farm and being a breeder of Suffolk horses. He must have remembered doing chores on cold snowy days because it was detailed in his design of the barn. Most chores could be done inside this barn. The barn lumber was salvaged from dismantling buildings on Black Road in Joliet. This barn was probably the last pegged barn built in Will County. The barn was built with the east side (from Donald Duck left) containing machine and equipment storage. Inside a small double door (behind sign) are stalls for horses on the left and stanchions for four milk cows on the right. The balance of the building (right side) has overhead storage for three thousand bushels of grain and storage for approximately eight thousand bushels of ear corn. This storage has recently been converted to shelled corn. About first window from the right was storage for ear corn which could drop into a burr mill below and be processed for cattle. The driveway has a scale to weigh wagons, trucks, and cattle from the lots. There is an inside elevator which goes up one side and down the other side with a dump in the floor. Both the scale and elevator are still used today. The barn is 105 feet long, 42 feet wide, 22 feet to the eaves, with a 28-by-85-foot cattle shed on the south side with a large concrete feedlot. The timbers under the crib bins are 10 feet by 12 feet by 42 feet, one piece. Rafters are 2.5 inches by 6 inches by 25 feet long and the barn still has its original roof. A large metal cupola was removed from the center of the barn in 1998. Our sister-in-law painted the pictures on the doors. Needless to say I'm an IHC owner. When the barn was finished, Senator Barr had a barn dance and party for everyone. Cars were parked along the road clear to Wilton Center (.75 mile). I was seven years old and never had so much pop and ice cream while my parents were dancing in the large haymow. My wife and I purchased the farm in 1969 from Senator Barr's family. Orin Porter, my great-grandmother's father, was the original owner of this farm. The farm is now a Centennial farm and will be in the family 150 years in 2004. This barn is located in Wilton Township near Wilton Center.

Photograph and information submitted by Burton Barr.

District 2, Boone County

Frosted trees form a halo over this barn located south of Belvidere. Sonny Sellers was the owner of this barn. Since this picture was taken the barn has been torn down.

Photograph and information submitted by Mildred Banks.

District 2, Cook County

William and Edith White purchased the farmstead with the brown barn in 1950. Their grandson, David Pennacchia, has helped with the farm chores including baling hay. The barn is located on Algonquin River Road, Barrington Hills.

Photograph and information submitted by David Pennacchia.

District 2, Kane County

This round barn is located near Pingree Grove. The views below show the interior structure of an attached portion and the wooden silo running through the center of the main barn.

Photograph submitted by Bill Clark without information.

District 2, La Salle County

This farm is known as the Eagle Farm. The barn, located on Bottom Road between Seneca and Marseilles, is beginning to be refurbished.

Photograph and information submitted by Paul A. Phelan.

District 2, McLean County

This was the first round-roofed barn to be built in the Saybrook area of McLean County. Herman Warsaw had this large round-roofed barn with its attached milk-cooling house built on his farm by Ebon Taylor in 1948. Contractor Taylor is shown in the doorway of the barn upon its completion. Mr. Taylor, who passed away at age ninety-five in 1994, was a builder by trade and built numerous barns and corncribs in the area.

Photograph and information submitted by Pat Evans.

District 2, Kendall County

Unique bent rafters (shown in small photo below) are around the top of this dairy barn, built in 1929. There must be hay and or straw in the haymow at all times to hold the barn together. Russell Wollenweber is the owner of this barn located northeast of Newark.

Photograph and information submitted by Joe Steffen.

District 2, Kendall County

This barn was built in 2002. It is a dairy cattle freestanding barn that holds up to 108 dairy cows. It measures 63 by 180 feet. Kirk Friestrad is the owner of this barn located near Central, at the intersection of Route 47 and Route 52.

Photograph and information submitted by Joe Steffen.

District 2, De Kalb County

This replacement barn, located one mile east of Cortland, was built in 1929 after the original horse and cattle barn, built in 1847, had been struck by lightning. The original settlers were the Rudolphus Burr family. The farm is now a sixth-generation-owned Centennial Farm. The 1929 barn was built with stanchions for twenty cows, a pen for five calves, a box stall for the bull, and six horse stalls. There was an oats bin, a ground corn bin, and a room attached to the silo. A driveway ran the length of the barn. Sliding doors separated the horses from the cattle. A stairway led to the haymow with a trap door at the top. Milk was shipped from this farm from 1931 to 1953.

Photograph and information submitted by Robert Jon Benson.

District 2, Kendall County

This barn, located near Newark, was built in 1840 by two woodchoppers from North Carolina who used logs taken from the nearby Fox River Valley Timber. It was the first barn built in Kendall County west of the Fox River. The burr oak beams are forty-two feet long without splicing. The outside timbers are black walnut. Chief Shabbona and his wives were known to be friends of the original owners. When they visited, their horses were kept in the barn. The hayloft is now a museum for natural history collections and Native American artifacts found in the area. The lower level of the barn is the home of the owners, the Howard Christensens. The silo is used as a lookout tower. Photograph and information submitted by Joe Steffen.

District 2, Livingston County

This barn was built in 1880 on property originally owned by John Wyllie who settled on the farm in 1866. John Hargneaues Wyllie, his great-grandson, is the current owner. The barn was built with lumber sent by rail from Chicago to Emington. The barn is located near Emington. Photograph and information submitted by Mary Capouch and Bonnie Reed.

District 2, Will County

This round barn with a sixty-foot diameter was built in 1898 and functioned as a dairy barn. The second level, which was accessed by way of a ramp, was used for hay storage. This is the second round barn at the Round Barn Farm. This barn was recently given a facelift that included straightening support beans, replacing siding, a new roof, and a fresh coat of paint. The barn is located at Manhattan.

Photograph and information submitted by Frank and Dolores Koren, Round Barn Farm and Museum.

District 2, Will County

John Baker built this multisided barn in 1898 on land homesteaded by his parents in 1850. The lumber used in its construction came from the Chicago Columbian World Exposition, which was being dismantled. The barn is sixty feet in diameter and contains over twenty stalls that were used for shorthorn cattle and workhorses. The second level was used for hay storage. The third level had several grain bins. Currently the round barn farm, located at Manhattan, is open to the public for educational and recreational purposes.

Photograph and information submitted by Frank and Dolores Koren of the Round Barn Farm and Museum.

District 2, La Salle County

The Maierhofer barn, built in 1899 one mile north of Seneca, stands as a monument to twentieth-century farming. It has been utilized by five generations of the Maierhofer family. The sixth generation will soon begin their FFA and 4-H projects, under its roof.

Photograph and information submitted by Jeff Maierhofer and Jim Kraft.

District 3, Adams County

This barn was built in the late 1800s or early 1900s. It was built in sections to accommodate horses, cows, and swine. It was used in the dairy business until about 1940 or 1950. The barn is located one mile east of the Quincy Airport.

Photograph submitted by Henry Bockhold with information by Michael Bockhold.

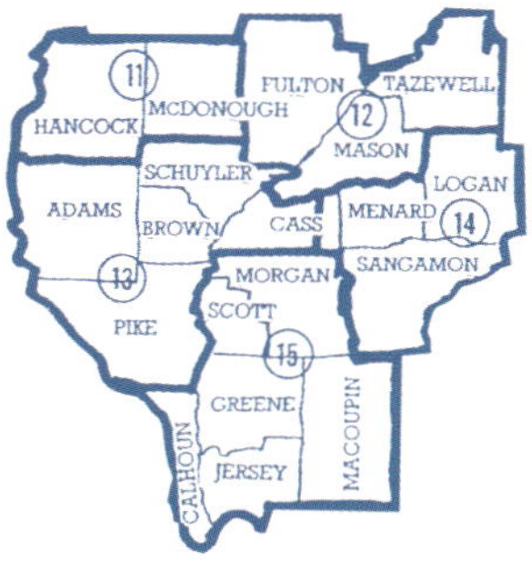

DISTRICT 3

The remaining barns of District 3 reflect the many different types of Illinois agriculture and its resultant prosperity. The dairy barns of Adams County reflect a once-active milk production region. The Victorian, cross-gabled barns of McDonough and Fulton Counties were built by well-to-do grain farmers to house their teams of draft horses, hay, and grain. Round barns were generally built to accommodate dairy operations, but they remain the jeweled icons of yesterday's agriculture. Basement barns of all types, including Pennsylvania-German basement barns, were built into hillsides where the lay of the land would allow. Gambrel-roof barns accommodated greater storage of hay and reflected growing livestock specialization.

By far the most used building material is wood. Local-quarried stone or fieldstone was used for the earliest basement barns. Over timber-frame structures, boards were nailed vertically side-by-side, sometimes overlapping in a method known as board and batten, to protect the interior from rain, sleet, and snow.

On the barns built by some of the more prosperous landowners decorative shingles, shutters, overhangs, and special gable treatments combine with the overall design of the barn to add to its visibility.

—Kent Slater
1963–64 State FFA President

District 3, Adams County

Richard Gray built this barn in 1871. This three-story bank barn measures forty by sixty feet. This barn, in the past, accommodated a large amount of livestock, hay, straw, and grain. The current owners are making efforts to restore it to its original condition. This barn is located at the north edge of Coatzburg.

Photograph and information submitted by Steve and Tina Keppner.

District 3, Morgan County

Oliver Cully built this three-story barn five miles north of Jacksonville in 1888. The barn is unusual in that it has two stairways. It was built to house up to twenty-two horses and four carriages. Several tons of hay could be stored in its large haymow. The barn was designed so grain could be moved by gravity.

Photograph and information submitted by Isabel Cully Norfleet.

District 3, Macoupin County

This barn has recently being going through renovation. It is located west at the exit off of I-55 onto Route 108 two miles, then north toward Atwater.

Photograph submitted by Paul Phalen.

District 3, Pike County

Thomas Barton, the original owner, purchased this farm in 1826. Elza Barton, a third-generation family member, built this barn in the late 1800s. The frame of this barn is joined together with wood pegs. The foundation is made of fieldstones and the walls are board and batten siding. The barn features a full basement. Kay Taliaferro, daughter of Harris and Lois Barton Franklin, is a sixth-generation owner. This barn is located on Old Barton Road near Pleasant Hill.
Photograph and information submitted by Kay Taliaferro and Harris Franklin.

District 3, Adams County

This barn was built in the early 1900s near Camp Point. It formerly housed dairy cows and pigs. In the summer months the barn is fully utilized as five hundred to six hundred bales of straw are raised by hayfork to fill the haymow. The current owners are H. and J. Bockhold.
Photograph submitted by Henry Bockhold and information submitted by Michael Bockhold.

District 3, Calhoun County

This older barn has stood for over 110 years. The small village of Damiansville was built around the farm. The barn, which once housed dairy cattle, had its hayloft converted to a basketball court so school children would have a place to play. Chickens were raised in the barn after it was no longer used for dairy cattle. In recent years the barn has been used for hay storage. This barn is located in the middle of Damiansville.

Photograph and information submitted by Henry Albers.

District 3, Menard County

This hip-roof barn, located six miles northwest of Petersburg, was built circa 1900. Originally dairy cows were milked in the basement portion. In the middle of the 1950s the basement portion was converted to a hog nursery with a pit, slatted floors, and sixteen pens. A grinder was used to mix the feed, which was then elevated to a storage bin on the second level where it could be augured out. The hayloft when empty was used as a basketball court.

Photograph and information submitted by Glenn, Steve, and Larry Fricke.

District 3, Sangamon County

This barn was built near Mechanicsburg, circa 1890. The additions were added circa 1940 and circa 1950. This barn was used primarily as a dairy barn but was also used for other farm animals as well. It is currently owned by Harry and Loretta Klaus.

Photograph and information submitted by Loretta Klaus and Margaret Klaus.

Barns of Pike County

District 3, Pike County
This round barn, located west of Pittsfield, makes a nice picture sitting in a field by itself. Photograph submitted by Mary Barnes.

District 3, Pike County
This barn and its companion buildings, located north of Detroit, make a great old-time farm scene.
Photograph submitted by Mary Barnes.

District 3, Pike County
A freshly cut wheat field surrounds this red barn with its white roof. The barn now serves its owner as a machine storage shed. Photograph submitted by Mary Barnes.

District 3, Pike County

This red barn with a white roof makes a great picture. It is located south of Barry. Photograph submitted by Mary Barnes.

District 3, Pike County

This white barn, located south of Griggsville has a dormer on the front and wood vents in the cupola.
Photograph submitted by Mary Barnes.

District 3, Pike County

This red barn, located south of Griggsville at the end of a white rock driveway, was built in 1865.
Photograph submitted by Mary Barnes.

District 3, Sangamon County

This barn is included on the Illinois Historic Preservation List. The Italianto barn, a rare type for Illinois, is a square structure with bracketed eaves. This particular barn is found north of Chatham on Route 4.

Photograph submitted by Association of Illinois Cooperatives.

District 3, Green County

This barn is located south of Carrolton. Partners Garry Pregler and David Jeffery are working to restore as much of the structure to its original condition as possible. The land was originally settled by the Parker family in the 1830s. The barn was built prior to 1859, but the exact date is unknown. The home, built after the barn, was used as a stagecoach house on the Springfield to St. Louis route. An adjacent carriage house is still standing on the property. The barn was built as a labor-saving structure, as carts loaded with grain could back into the basement and unload grain. The present-day stalls were originally used as corncribs, storing crops of corn, oats, and other grains. Hay chutes were also built into the structure, making the transfer of bales or loose hay much easier from the top floor to the bottom. The two partners plan to have the barn restored in five years, depending on time and money. They say it takes one day to tear down a structure but it takes years to restore them. This is a good example of private owners trying to restore a historical structure. This barn appeared in the October 2000 issue of the Illinois Country Living *magazine.*

Photograph submitted by Association of Illinois Cooperatives.

District 3, Sangamon County

John W. Taft and his sons built this thirty-by-eighty-foot horse barn in 1950, using lumber cut and milled from timber on the farm. Originally the barn was built to house Percheron draft horses. Today the barn is used to house saddle horses and show cattle. The barn is located two miles south of Mechanicsburg.

Photograph and information submitted by Matt and Lynn Buckles.

District 3, Sangamon County

This drive-in two-story barn is included in the 1874 publication of the Sangamon County Atlas. It is located one mile east of Mechanicsburg. J. N. Fullinwider built the barn of hand-hewn logs and white pine siding. The manger and feed bunks stretched from one end of the barn to the other in the basement area. The ground floor has four large granaries with trap doors. Feed and grain could be dropped down into the mangers and feed bunks below. The barn housed a few milk cows, however its main use was to house jacks for breeding. In 1994 the barn was covered with white metal siding.

Photograph and information submitted by Matt and Lynn Buckles.

District 3, Sangamon County

This masonry barn was built just before World War II, for use as a dairy barn. This barn originally had a companion silo. The barn features concrete feed bunks, six milking stanchions, a milk-holding tank, a large loafing area, and an office. Due to the ruggedness of the brick structure, the barn is currently used in the winter to house bulls. The barn is located on the west edge of Mechanicsburg.

Photograph and information submitted by Matt and Lynn Buckles.

District 3, Adams County

This round barn was built in 1814 near Mendon. In the middle stands an eighteen-by-forty-foot silo of fir lumber. The frame is native lumber taken from trees on the original site and sawed into lumber there. The Lewis family donated this barn to the Adams County Olde Tyme Association, Inc. The roof and sidewalls were taken down in sections. It was moved fifteen miles and rebuilt on a concrete foundation near Camp Point at the Adams County Fair Grounds. The barn will now serve as a museum for all to enjoy. There are two barns at the same location.
Photograph and information submitted by Max Dickhut.

This forty-by-sixty-foot peg barn was built in the early 1900s. Pat and Kenneth Cramm donated this barn to the Adams County Olde Tyme Association, Inc. Thirty workers dismantled this barn and later reassembled it and added the basement. The barn is located at the Adams County Fair Grounds.
Photograph and information submitted by Max Dickhut.

District 3, McDonough County
The J. W. Everly barn was built in 1882. It is thirty by sixty feet and stands forty feet in height. It is presently owned by the L. W. Early Trust, which is administered by the Citizens National Bank of Macomb. This barn is located west of Adair. Photograph and information submitted by Kent Slater.

District 3, McDonough County
Leo and Mary King's cross-gable barn was built by Newt Willis in 1898 for James Forrester. Forrester used the barn for horses and hay. Even though Mr. and Mrs. King used it for hogs, cattle, and hay, they always referred too it as "our horse barn." It is in an excellent state of repair. This barn is located south of Walnut Grove.
Photograph and information submitted by Kent Slater.

District 3, Fulton County

This barn was built by Sylvester Martin near Havana. The exact year is unknown. The current owners are Daryl and Susan Cheney. The barn was part of a feedlot and the cows were fed on the sides with hay stored in the middle part. There are still stalls on the sides for cattle. Today, the barn is only used for storage.

Photograph and information submitted by Scott E. Cheney.

District 3, Macoupin County

This barn, located near Brighton, was built in 1835. Allen E. Tucker's family has owned this property since 1870. The frame has hand-hewn oak mortise and tenon and pegged construction. Square-cut nails were used. Some of the original siding boards are up to thirty-six inches wide. The original foundation consisted of round logs standing on end. Mr. Tucker's great-grandfather put in rock piers, which can still be seen. The rocks were hauled from Alton, a distance of fifteen miles. This barn has always housed cattle and at one time it had six horse stalls. Eighty-year-old Mr. Tucker has lived on this farm his entire life. He still remains active working in the barn with his cattle.

Photograph and information submitted by Allen E. Tucker.

District 3, Fulton County

This barn is about sixty-five years old. In 1998 it needed a facelift. White paint was being applied when this photograph was taken. With its new roof the barn is as good as new. Originally constructed to house horses and dairy cows, it is currently being used for Black Angus cattle and a small feeder pig operation.

Photograph and information submitted by Robert H. Rector.

District 3, Fulton County

Charlyn and Allen Martin are the owners of this barn located on County Highway 39 near Farmington. Charlyn remembers in the 1930s and 1940s driving teams of horses and or mules to pull hay up into this barn. This barn has been updated in recent years with siding. Following a fire in one end, the roof was replaced.

Photograph and information submitted by Charlyn Martin.

District 3, Fulton County

This barn stands in a field located between Lewiston and Bernadotta. It is still tall, straight, and plumb in spite of having lost a large portion of its walls. It is a credit to its builder.

Photograph and information submitted by Charlyn Martin.

District 3, McDonough County

This impressive cross-gabled barn was built in 1909, by Nick Breasaw of Busnell, on the H. D. Lantz farm south of Walnut Grove. Mr. Breasaw modified and expanded on the Willis cross-gable design by enhancing the size of the middle ridge row, by adding architectural detail, and by recreating the roof design in the cupola. Originally built to accommodate horses, hay, and grain, this barn was later converted to a dairy operation. It was torn down in the late 1980s. This barn was located 1.5 miles north of Bushnell.

Photograph and information submitted by Kent Slater.

District 3, Sangamon County

This blue round barn, located near Glenarm, is now used as a hog barn. It may well be one of the few blue barns, of any shape, in Illinois.

Photograph and information submitted by Lois Herr.

District 3, Morgan County

This barn, located three miles north of Franklin, was raised in the late 1800s. It was constructed with wooden pegs. It has been used, maintained, and preserved by three generations of the same family. Photograph and information submitted by Mrs. Geneva Sweet.

District 3, Hancock County

The age of this barn, located near Sutter in Walker Township, is unknown. The current owners, Lance and Layne Roskamp, purchased it in 1979. The barn has a large hayloft and two granaries. Photograph and information submitted by Jami Roskamp.

District 3, Fulton County

These two old large barns are no longer standing but were great structures in their day. These two large barns were destroyed in the 1990s. The barn on the left was a cattle barn and the barn on the right was a horse barn. The Alcott family built both of these barns prior to 1900 near Farmington.

Photographs and information submitted by Charlyn Martin and Kent Slater.

District 3, Fulton County

This horse barn, with its porch on the front for storing machinery, was moved in the early 1900s. The barn had glass windows with bars across them on the east and west side of the lower level and the upper level. The horses' heads were faced to the outside walls in individual stalls. Feeding was done from the upper level by chutes for grain and hay. The barn was used to house racehorses and had an exercise track on the west side of the barn. This barn was located on Markly Road, near Middlegrove. Photograph and information submitted by Charlyn Martin.

District 3, Adams County

Basement barns are a predominant style in the Quincy area and this barn located on Route 96 north of Quincy is typical of the style. John and Margaret Alison are the proud owners of this barn and utilized it until 1986 for their Holstein cattle and Shropshire sheep.

Photograph and information submitted by Kent Slater.

District 4, Piatt County

This barn sits high on a working farm just west of the University of Illinois, Allerton Park. Photograph and information submitted by Paige Buck.

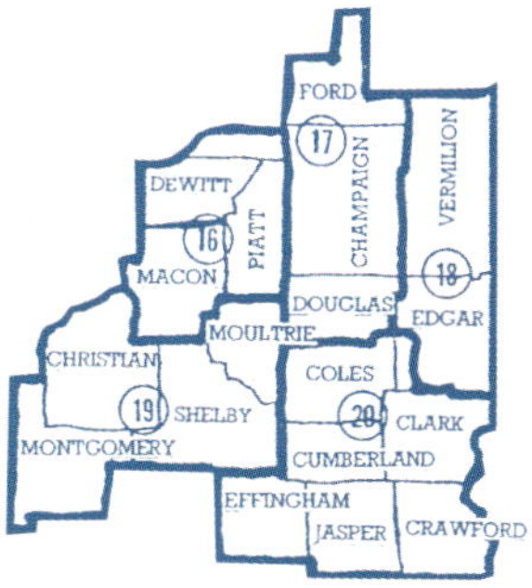

DISTRICT 4

District 4 includes some of the most fertile land in the state. Some of the land was originally swampland as shown on early maps of the area. But the arrival of dredging machinery that dug manmade ditches, feeding the nearby rivers, and the installation of many miles of drainage tile, made those areas of swampland into fertile farmland.

Elsewhere, the many years of prairie-grass growth made the land very difficult to till. But the John Deere plow changed all that. But, even with the new plow, the soil required oxen to pull the plows through the rock-like crust for the first time.

The arrival of the railroad across central Illinois made it possible to reach the Eastern market with farm crops. This brought settlers from many different backgrounds and cultures, and small communities sprung up along the railroad routes. The Great Western Railroad crossed Piatt County, and Ivesdale, Bement, and Milmine were among those communities.

Settlers in District 4 included immigrants from England, Ireland, and Sweden, as well as others. A majority of them were farmers. They descended on the Danville Land Office to purchase these lands. Acreage located adjacent to the new railroads was slightly more expensive, and the prices were somewhat less, depending on distance to the railroad.

This influx of settlers caused the construction of numerous barns all across the area, and, depending upon the different heritages, a wide variety of elegant barns were erected. For many years these barns stood as eloquent testimony to the farm industry and its importance to this area.

But, as farm methods changed and automated farm machinery replaced the horse-drawn varieties, the barns lost their important place on the farmsteads of the area. These wonderful rural icons in this area are fading fast, and we must enjoy to the fullest those that are still around while we can!

—Robert E. Bodman

District 4, Ford County

This barn was built near Piper City in 1902. It is one hundred feet long and forty-five feet wide. In the haymow there are two large grain bins. The barn had a large cupola with a large weathervane on top that blew off sometime back. Mrs. Dietz's husband used to tell of climbing up into the cupola and catching squab to eat. Mrs. Dietz still has the original floorplan of the barn. When originally built the barn had stalls for twenty-eight horses and ten milk cows and calves. In 1947 it was remodeled for dairy cows entirely. It is now being used for pigs. Photograph and information submitted by Nelda W. Dietz.

District 4, Macon County

Frank Brown built this peg-constructed horse barn circa 1910. The current owners are Randal and Beth Fransen. They have owned this piece of Illinois history since 1991. The barn is located four and one half miles west of Booty.

Photograph and information submitted by Randal Fransen.

District 4, Clark County

Hayden Daugherty, a Civil War veteran, along with Joe Byram built this log barn near Martinsville in 1865. Hayden Daugherty was the great-uncle of James C. Daugherty's great-grandfather of rural Martinsville. The barn was torn down in 2001.

Photograph and information submitted by submitted by James C. Daugherty.

District 4, Piatt County

Nels Larson, pictured by the door below, was known as the "Squire of Voorhies." In 1904, he built this Swedish barn with its four-faced, sixty-eight-foot-high clock tower for his driving teams and carriages, in spite of the fact he did not like horses. The Seth Thomas clock weighed three thousand pounds. In 1976, the barn and the companion tilting windmill were destroyed by a tornado. Mr. Larson's home, the refurbished Voorhies Castle, still stands and is a private residence. The current owner, Steve Seitz, has plans to rebuild the clock tower barn. Voorhies is located four miles south of Bement. The large photo to the right was taken shortly after the barn was built. The photo above is one of the last pictures taken of the barn.

Photographs submitted by Duane and Martha Medaris and Bill Clark. Information provided from records of the Piatt County Historical Society.

District 4, Effingham County

This barn, located north of Effingham, is over one hundred years old. It was built with pinned construction. It has split rail rafters and originally had a wood shingled roof. Its unique design allows for inside unloading of hay.

Photograph and information submitted by Leland Huelskoetter.

District 4, Piatt County

This barn is believed to be the oldest standing barn in Piatt County that is still used as a barn. Henry Larson is the owner. His father, Lambert Larson, purchased the farm in 1906. At the time of the sale, Mr. Larson was told that the barn was at least sixty years old. This barn was used along with two companion barns to feed and house cattle. Cattle were shipped by rail. After traveling long distances they were unloaded and driven to this farm to be watered, fed, and rested up before continuing their journey. The other two barns were struck by lightning and burned the same night. A unique feature of the barn is the door within a door. The addition on the right was added in the 1950s. The barn is maintained in excellent condition and is used to house a few feeder cattle. This barn was featured on the 2002 Piatt County Barn Tour. The barn is located three miles south of Bement.

Photograph provided by the author. Information provided by Henry B. Larson.

District 4, Champaign County

These barns have been in Shelley Libka Place's family for three generations. Her grandfather, Lon Libka, built these barns around 1910. He would buy local livestock around the area and then rent a train car to ship them to Chicago to sell. Auctioneers would hold auctions right in the barn. The sign reads, "Lon Libka, Prop. Salt Fork Stock Exchange." My father, Leo Libka, owned and used the barns until he passed away. He raised cattle and hay. The barns were always filled with cattle and hay when I was growing up. The Hooker brothers from Sidell and Indianola area helped restore the barns and now I plan to use them. It took twenty-six gallons of red paint and ten gallons of white paint. The barn has never been spray painted, always hand brushed. The barn is 120 by 32 feet. This barn is located one mile north and one mile east of Homer.

Photograph and information submitted by Shelley Libka Place.

District 4, Champaign County

This barn, located one mile west of Homer, was built in 1935 by the landowner Willis Loyd. It was built and used as a cow-milking barn. The Gerald Compton family purchased the farm in 1950. It was used as a cattle barn for several years. In recent years it was remodeled and is being used to house an antique truck and for farm machinery.

Photograph and information submitted by Gerald and Ruth Compton.

District 4, Jasper County

Warren E. Harding's grandfather built this barn in 1895 for horses and mules. The barn, located south of Ste. Marie, has a full hayloft with a hay track running the full length of the barn. The hayfork is still on the track. The sloping addition was added in 1910 to serve as a wagon shed and a granary.

Photograph and information submitted by Warren E. Harding.

District 4, Piatt County

This old horse barn is the childhood barn of the author, it was built around 1887 by Nels Larson. It was designed after a barn that was near Nels Larson's childhood home in Sweden. The rounded windows and extra height to the hayloft are trademarks. The barn was originally set on field rocks. In the 1920s, the foundation was concreted, leaving some of the rocks visible today. This barn was built prior to the famous clock tower barn at Voorhies. All the tenant farmers that used this barn all had the first name of William. This barn was on the Piatt County Barn Tours of 2002 and 1998.

This picture shows Brodie Gallagher playing in the grain bin. The author says it brings back memories of when he played in the same grain bin. Brodie is the son of William Gallagher, the current user of the barn.

This picture shows the windows designed after a barn in Sweden. This barn is located 1.25 miles east and .50 mile south of Voorhies. Photographs and information submitted by Harry Porter.

District 4, Vermillion County

William and Rosella Ray are the owners of this barn located five miles east of Rossville. They purchased the farm in 1972 from Francis J. Christman whose family had owned the farm since 1876. This barn is reportedly the oldest three-story barn in Vermillion County and the second largest in the area, being one hundred by sixty feet, with twenty-eight-foot corner posts of hand-hewn lumber with wooden-peg construction. There are stalls in the basement for twenty-four head of horses and room for two carloads of cattle. On the second floor, there is room to store implements in addition to bins for grain. The haymow will hold one hundred tons of hay comfortably. The barn was constructed with hand-hewn lumber and fastened together with wooden pegs. The barn contains stalls for horses, a cattle shed, implement storage, and grain bins. The barn is still in use today.

Photograph and information submitted by William and Rosella Ray.

District 4, Clark County

This barn was built in 1922 by Kris Johnson's great-grandfather. The barn, located northest of Annapolis, was used by a great-uncle in his sawmill operation. Despite its age, the barn still presents a happy face.

Photograph and information submitted by Kris Johnson.

District 4, Douglas County

This barn near Tuscola was built in 1912 by Douglas R. Williamson's grandfather who is shown here with his faithful companion. The barn is built with heavy rough-sawed white oak framing and is mortised together with homemade wooden pegs. The barn is currently being used to house sheep. Photograph and information submitted by Douglas R. Williamson.

District 4, Piatt County

This barn was built in 1911. It was constructed with oak lumber hauled by horse and wagon from a sawmill located four miles north of the farm. The beams are held together by peg and mortise tenon. It was originally built as a horse barn and had eight single stalls and three box stalls. In the 1930s, two of the stalls were converted into a two-car garage. In the 1950s it was converted into a hog barn. Now it is used only for storage and a lot of good memories. The barn is located just northeast of Pierson.
Photograph and information submitted by Robert and Shirley Scheffer.

District 4, Crawford County

This barn is located on 375th Road near Flat Rock. The red barn with its large hay hood is well maintained as you can see by this aerial photograph. The barn was built in 1906 at a cost of $400. Using native lumber, it was constructed with wooden pegs and fasteners. Photograph and information submitted by Owen Edgington.

District 4, Champaign County

This barn, located near Champaign, has stood for over one hundred years according to its owner Vince Early. A few years ago, Mr. Early debated about whether or not to tear down the old barn and build a new shed. Instead he elected to repair it. He has been happy with his decision.

Photograph and information submitted by Vince Early.

District 4, De Witt County

The year this barn was built is unknown. The 1956 Ford automobile was placed on top of the barn in the late nineteen fifties. The story goes that the then owner of the barn Mr. Legget had a daughter whose car had engine problems. The cupola was blown off the roof by high winds. Mr. Leggett used a block and tackle with blocks on the roof. He raised the car from the inside of the barn and set it on the roof to serve as a cupola—no small feat. The barn is located about two miles north of Wapella.

Photograph submitted by Anne and Bob Steinman. Information submitted by Dean Stevens.

District 4, Piatt County

Robert Allerton owned over a thousand acres of farmland in Piatt County. This barn was near the farm manager's house and was used as a cattle barn. The barn is unique in that it has a metal spiral stairs to the large hayloft. A winter and summer scene are shown. This barn is located at the south entrance of Allerton Park near Monticello.

Photograph submitted by author. Information submitted by David Bowman, manager of Allerton Park.

District 4, Ford County

This neat white barn with its wood shingled roof and concrete foundation is in good repair. This style of barn can be found throughout Illinois. The barn is located west of Gibson City.
Photograph submitted by Paul Phalen.

District 4, Champaign County

This red barn with its white doors was built in 1914. It is sixty by sixty feet with oak beams mortised together with oak pins. The exterior is original. It has a concrete floor, two staircases to the haymow, a crib, and box stalls. The barn is located between Mahomet and Champaign.
Photograph and information submitted by Thelma and Lloyde Primmer.

District 4, Champaign County

This barn was built in 1915 and constructed of cement blocks, manufactured one at a time by a "Block Machine" which is still in the possession of the owner. The material used is native gravel. The house was also built in the same manner. The barn was built state of the art for its time to accommodate draft horses used in gravel work business. The barn is located near Fisher. Photograph and information submitted by Ron Cook.

District 4, Clark County

This barn has been converted into a restaurant and is called Richards Farm Restaurant. It is a popular dining place just off of Route 40 at Casey. The barn in addition to being a restaurant also features a garden area for all to enjoy. The food is good and it's a nice place to visit. Photograph submitted by Paul Phalen.

District 4, Moultrie County

This barn, located south of Arthur, is on an Amish farm and still in use today.

Photograph and information submitted by Jo Ellen Seaman and the Arthur FFA Chapter.

District 4, Clark County

Arthur and Mattie Guinnip had this dairy barn and wooden silo built in 1926. The milk from a herd of Jersey cows was bottled and delivered in Marshall. This barn is located near Marshall.

Photograph and information submitted by Donald Guinnip.

District 4, Moultrie County

This Amish farm is located south of Arthur. This barn at one time housed purebred Jersey cattle. When the disposal sale was held, the Wrigley's from Chicago came down and purchased some of the Jersey cows.

Photograph and information submitted by Jo Ellen Seaman and the Arthur FFA Chapter.

District 4, Montgomery County

This old horse barn was built on the Walter Zimmerman farm approximately four miles southwest of Morrisonville. The barn, although built as a horse barn, was mainly used for cattle, although throughout the years it has been used for sheep and cattle. It is shown here all decked out with its snowy roof and the frost on the trees on an early winter morning. Notice the hole in the door for the family pets to go to and from. Photograph submitted by Katie Weldeck.

District 4, Douglas County

This red round barn is located near Filson. No other information available on this barn. Photograph submitted by Bill Clark.

District 4, Champaign County

Roy Stout owned this barn, until he sold it, and it was converted into a restaurant. The barn was then known as the Round Barn and Convention Center. The barn is now occupied by Famous Dave's BBQ Restaurant. This barn is located at the corner of South Mattis Street and Springfield Avenue in Champaign.

Photograph submitted by Paul Phalen.

District 4, Coles County

This barn was built in 1916. Known as Halsey's barn, it is located on Route 130 east of Charleston.

Photograph submitted without information by Lois Herr.

District 4, Cumberland County

This round red barn with its white roof and white trim is located in Cumberland County southeast of Toledo. The milk-cooling house is now used as a garage.

Photograph submitted by Paul Phelan.

District 4, Ford County

This weathered old barn, located west of Gibson City has served its owner well over the years but now stands near the Osage Orange trees planted after it was built. It is slowly losing its weather boards but still stands as straight as when it was built.

Photograph submitted by Paul Phalen.

District 4, Champaign County

This barn was built about 1915. Wanda Taylor Legg's grandfather, David Andrew Taylor, originally purchased the farm. Robert Reiners is the third generation of his family to farm the land. The barn, demolished in 1996, was located southwest of Gibson City.

Photograph and information submitted by Mary Capouch.

District 5, Pulaski County

This barn, located near Mounds at Pearl Acres, is owned by the Pearl family. No other information available. Photograph and information submitted by Kathy Brey.

DISTRICT 5

Easily accessible by rivers, southern Illinois was settled before other parts of the state. Later settlers came to the area mainly from the southeastern states. Early commercial industries included salt, iron, coal, forest products, and agriculture.

The region is often referred to as Little Egypt. One version of this label is that the sparse crops in areas to the north sent farmers south to purchase corn for their livestock. They likened their plight to the sons of Jacob, "they were going down to Egypt for corn."

The landscape of the region changed through time from being a very diversified farming region (grain, livestock, vegetables, and fruit) to larger farms raising grain (corn, soybeans, and wheat), cattle, and hogs. Part-time farms and rural residences are now an important feature of the landscape.

A different type of diversification is returning to the area as wineries and fish farming develop. A large portion of the area formerly in farms and in strip mines is now in the public domain (Shawnee National Forest, wildlife refuges, and state parks).

—Lois T. Herr

District 5, Perry County

This barn, located near Pinkneyville, was the dream of teenager William B. "Pop" Timpner. His father, Edward, provided financial support and guidance to make the dream a reality. Construction took place from 1916 to 1918. This is the largest barn in Perry County. The same family has owned this farm for over 130 years. The family still holds dances in the spring before hay production starts. Photograph submitted by Ken Kashian. Information provided by The National Trust For Historic Preservation, Barn Again Program.

District 5, Randolph County

John Surman built this barn in circa 1920 with the help of his neighbors using large logs, which were squared off. Limestone rocks were used for the foundation. When the barn floor was completed a dance was held. The current owner, Shirley Dunker, is putting on siding. The barn is located near Chester.

Photograph and information submitted by Shirley Dunker.

District 5, Randolph County

Clarence Tegeler built this barn located near Chester in 1960s. It was built from oak trees cut on the farm and sawed into lumber. The dimensions are eighty by one hundred feet. At the time it was built it was thought to be the largest barn in Randolph County.

Photograph and information submitted by Shirley Dunker.

District 5, Richland County

Of interest to travelers on Route 50, this barn has a large advertisement for Mail Pouch Tobacco painted on the north side, protected from the scorching sun. The barn is located east of Olney.

Photograph and information submitted by Roselyn Nesbit.

District 5, Union County

Carol Hanson is the owner of this 130-year-old mule barn located at Dongola. It is constructed of popular and oak with wooden pegs. Originally it supplied mules to area farmers, to the logging industry and to the U.S. Army. Since that time it has served as a car dealership, a lumberyard, and currently houses the Fragrant Fields Nursery and Garden Center and the Mule Barn Cafe and Gift Shop.

Photograph and information submitted by Lois Herr and Ken Kashian.

District 5, Union County

This barn, built circa 1917, is owned by Robert Kimber and is shown here with its reflection in the farm pond. Mr. Kimber is a direct descendant of the Kimber family that settled in southern Illinois in the 1840s. This barn has mortise and tenon construction with wooden pegs. The barn features a gambrel roof. A unique feature is the wide roof overhang that completely circles the barn, which makes a covered walkway around the barn much like a porch. It housed workhorses and dairy cattle and later was converted to housing beef cattle. The barn is located near Dongola.

Photograph and information submitted by Lois T. Herr.

District 5, Clinton County

This old log barn has a lean-to on each side and a hayloft. It is located on land purchased in 1885 by Jacob Jenne who had emigrated from Germany. Since that time the farm has been in the Jenne family and is currently owned by Jewel Jenne. The barn is located north and west of Carlyle.

Photograph and information submitted by Eldon Greenwood.

District 5, Wayne County

This barn is located near Cisne and is owned by friends of the person who submitted the picture. Photograph submitted by Karen Barnum.

District 5, St. Clair County

Early in the eighteenth century most of St. Clair County around Columbia was covered with a dense forest of hardwoods. The early German settlers that developed this area cleared the land and used the trees for their own benefit. One German family named Weilbacker developed a large area of land, which was owned by them for up to the fourth generation. The barn is located at Millstadt.

Photograph and information submitted by Mitlon Heitman.

District 5, Bond County

Matt Willmann built this barn in 1929. Mr. Willmann is the grandfather of the current owner, Arnold E. Hunter. Mr. Hunter's father, P. A. Hunter, used the barn as a dairy barn. The barn is located on Indian Creek Avenue near Greenville.

Photograph and information submitted by Arnold E. Hunter.

District 5, Jackson County

Clyde and Wilma Canada were the original owners of this log barn located on Kennedy Road near Carbondale. The barn was built in 1830 from sycamore logs harvested from next to the creek which traverses the property. The side overhangs contained stalls for horses and cattle, while the center portion was used to store hay. The farm produced vegetables, strawberries, and peonies. The current owner, Janneke Resnick, at one time raised rabbits for their fur. The barn is now used for storage and tools.

Photograph and information submitted by Lois Herr.

District 5, Union County

This horse barn is owned by Barbara Throgmorton. Her great-grandfather, James K. Walton, built it in 1864. That date was carved in one of the foundation stones, which were quarried on the property. Hand-hewn oak logs, some fifty feet in length, were used. Support beams, ten inches square were secured using mortise and tenon joinery. The barn currently houses sheep. Approximately forty lambs were born in February 2002 on this farm located near Anna.

Photograph and information submitted by Lois T. Herr and Kathy Brey.

District 5, Union County

This barn was built in 1945 with wooden pegs and mortise and tenon construction on a concrete foundation. Nancy Murray is the current owner of this barn located south of Anna.

Photograph and information submitted by Lois T. Herr.

District 5, Monroe County

John Meyer, a cabinetmaker from Waterloo, built this barn from 1917 to 1919 for its original owners, John and Christine Mueller (Miller). John Miller's father originally purchased the land in 1853 for $1.25 per acre. The barn is located near Waterloo.

Photograph and information submitted by Kathy Reime.

District 5, Franklin County

This barn was built by Leonard Trader from trees on his own land, circa 1920. The trees were cut with a crosscut saw. The sills were hand hewn. The farm, located near Sesser, has been in the same family for four generations.

Photograph and information submitted by Tom and Lucille Trader.

This barn was built near Sesser during the Depression, circa 1930. The hayloft has a hardwood floor in perfect condition. When first built it was the site of several barn dances.

Photograph and information submitted by Tom and Lucille Trader.

District 5, Williamson County

Wesley Trammel built this barn near Stonefort in the mid-1860s. He had settled in the area in the early 1800s, the current owners are the direct descendents of Wesley Trammel. The barn is constructed of oak beams and wooden pegs. The side wing was added later to the original barn.

Photograph and information submitted by Lois Herr.

District 5, Bond County

This barn stands tall on a hillside farm that the late Lawrence Rutz moved onto around 1932. Approximately two years later, the farm was bought by the late Dr. A. F. Kaeser of Highland, Illinois, and is still owned by the third generation of Kaeser's family. The operation of the farm, however, has remained in the Rutz family and currently Ronald Rutz, son of Lawrence, resides on the farm. Ronald, who was born on the farm in 1938, helped milk a herd of twenty-four dairy cows in this barn until 1968. Old-fashioned milk cans were used up until 1963 when a bulk storage tank was put in. The barn also has a hayloft that has housed numerous years of straw and hay and an adjacently attached milk house that is the "office" of the cattle operation. Currently the barn houses registered shorthorn beef cattle that the Rutz family has been raising for over twenty years. This barn is located at 4668 Fairview Road near Pocahontas.

Photograph and information submitted by Rene Mettler.

District 5, Bond County

This barn, located in Old Ripley Township near Greenville, was built in 1882. It has been a dairy barn, horse barn, loose hay barn, ear corncrib, straw barn, and a neighborhood party barn.

Photograph and information submitted by Marshall Mollet.

District 5, Randolph County

This barn, located two miles east of Ellis Grove, is over one hundred years old. It is unchanged except for a fresh coat of paint.

Photograph and information submitted by Lloyd Schoenbeck.

District 5, Jackson County

This barn represents one of the last "barn raising" events in Jackson County. It was built in 1940 on the sandstone foundation of the original barn, which burned down when struck by lightning. The lumber was precut, numbered for assembly, and pieced together with bolts. It has a metal roof and a hay track for loading hay. The barn has stanchions for milk cows. The barn was also used to house horses and chickens. Keith Lingle is the owner of this barn located near Carbondale.

Photograph and information submitted by Lois T. Herr.

District 5, Johnson County

Barns come in all sizes and shapes and have many different uses. While this barn on U.S. Route 45 was built to house livestock and store hay, it is now used to harvest the tourist trade.

Photograph and information submitted by Lois Herr.

District 5, Jackson County

This peppermint-striped barn was built in 1925. It was a roofed lean-to on all four sides and a matching striped cupola. The barn is in a beautiful setting located six miles south of Ava on Dry Hill Road.

Photograph and information submitted by June Black.

District 5, Union County

This barn was built in 1878 as indicated by the date carved in a foundation stone. It was constructed using hand-hewn white oak logs. It has mortise and tenon construction, with wooden pegs. This barn, owned by Philip Otrich, is located south of Anna.
Photograph and information submitted by Lois Herr.

District 5, Washington County

The oldest building in Okawville is this pre–Civil War barn built in the 1850s. Faint pencil writings on the inside walls recorded who borrowed grain from the original owner and builder. The next owner, Jacob Stricker, was a rural mail carrier. He utilized the barn to house the horses used on his mail route.

Photograph and information submitted by Warren and Virginia Stricker.

District 5, Saline County

This barn, located on the edge of Harrisburg, is known in the area as the "Mule Barn." The barn was built in the 1920s, and it survived the flood of 1927. Mr. Parker bought mules that were in poor condition from being used in the nearby coal mines. Through rest and good nutrition he restored them to good health and then sold the mules back to the mines. The mules were bred to be small and strong so they could work in the mines.

Photograph and information submitted by Lois Herr.

District 5, Pulaski County

Edward Lackey built this Sears and Roebuck barn in 1929 at a total cost of $987.00. In its early days it was a cattle barn and is now used for horses. It is constructed of pine. The original forms for making the curved-lathed rafters for the arched roof are stored in the barn. Note that the windows tip inwards so that the rain will not enter the building. The current owners are Eugene and Susan Parker. This barn is located south of Pulaski.

Photograph and information submitted by Lois Herr.

District 5, Johnson County

Frederick Heaton is the owner of this barn located near New Burnside. It is constructed of white oak and siding of cedar. A unique feature of this barn is an interior vertical elevator having a floor of about eight feet square. The elevator, located under the cupola, is used to access the three levels within the barn. It is operated by hand using a series of wheels and pulleys. At the time the barn was built in the 1900s the area was devoted to orchards. One use of the barn at that time was to store and mix pesticides as required by local orchardists. Photograph and information submitted by Lois Herr.

District 5, St. Clair County

This barn was constructed circa 1904 when the Consolidated Coal Company constructed its Mine No. 17, a shaft mine. The barn housed the mine mules when the mine wasn't working. It may also have served to stable the superintendent's horse. The superintendent, Peter Grieve, had a horse (Dolly) and buggy to travel to his home and to work at the mine. The mine closed in 1926 and the mine yard lay unused for several years. In 1937 William and Frances Bauer moved to the superintendent's house. He served as mine watchman. The Bauers got a milk cow, a pair of mules, and two milk goats. The barn served to house them. Soon the Bauers expanded to raise chickens, turkeys, and geese—none of which got to enjoy the barn, of course. Later Hereford beef cattle became the main farm animal activity. This continued until the 1980s when William Bauer could no longer care for the animals or deal with haymaking or straw baling. In 1987, after the cattle were sold and William went into a nursing home, his wife Frances started renting to racehorse owners. Fairmount Park at one time had both thoroughbred and harness racing. When the harness racing season began, the thoroughbreds had to vacate the track. Providing a place for them put the barn and pasture to good use. In the past few years, harness racing has ceased. Now the thoroughbreds can stay at the track all year. Frances has died, and her daughter, Doris M. Bauer, continues to live on the site. Occasionally, she rents it for housing disabled racehorses or for saddle horses. Rather then destroy it, she renovated the barn with a new set of rafters and a new metal roof. This barn is located on Bethel Road near Caseyville.

Photograph and information submitted by Doris M. Bauer.

District 5, Johnson County
This barn, located near Vienna, makes a great backdrop picture for a southern Illinois rural scene. Photograph submitted by Bill Clark.

District 3, Hancock County

The Whitaker Barn was a catalog barn built in 1920 and has a Gothic roof. The barn was originally located in St. Albans Township, Section 16. It was dismantled in 1995 and rebuilt at its present location in Carthage Township in 1996. It is used for family picnics and also for a family wedding.

Photograph and information submitted by Donald Bumphrey.

District 1, Henry County

The Yenkey family are the proud owners of this barn located north of Geneseo on Grange Road. The barn is over one hundred years old, and is located on the original Samual Alan farm. It has been in the Yenkey family since 1944. It originally had a sand floor, which is now concrete. The basement was made of concrete bricks, which have been replaced. The original wooden floor on the main level is still in place as of today.

Photograph and information submitted by Drew DeRiemacker.

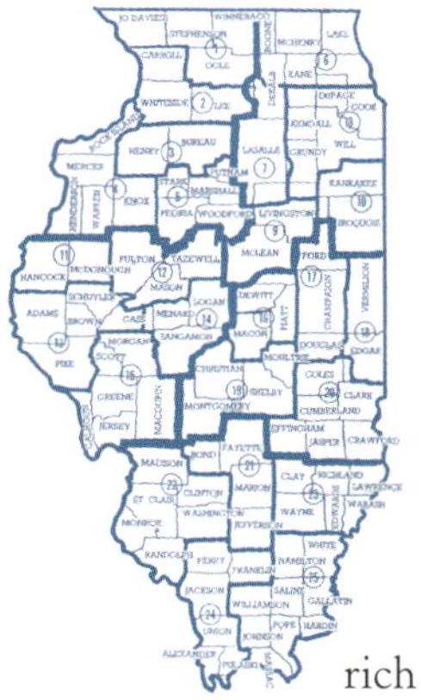

MORE GREAT BARNS OF ILLINOIS

Illinois is a state known primarily for its rich fertile farmland, diverse heritage, its large grain elevators located approximately five miles apart in almost checkerboard fashion throughout the state. The grain elevators were located along the railroad lines that crisscrossed the state from one end to the other. They were situated so that a team could pull wagons full of excess grain to market. The region is also noted for the agriculture and educational research work done at the University of Illinois and other great educational institutions located throughout the state.

Illinois was originally part of the Indiana territory until it became a state. The area was originally settled mostly along the waterways and high ground because of the thick prairie grass and lack of timber to build their barns in the thick mucky soil of the flat prairie.

The arrival of the dredging machines, that dug man-made streams to the rivers of the district, and the installing of drainage tile, opened up huge amounts of acreage that was not farmable before.

With the arrival of the railroads small communities sprung up along the rail lines. The need for milk, meat, and produce resulted in more land being settled and cultivated by farmers. Milk trains became the farmers transportation system hauling their produce to larger cities. The arrival of the railroad brought more people of diverse heritage to the region such as Irish, English, French, Norwegians, Germans, and Swedes. The mingling of different heritages brought a diverse style to barns built in this region, with the primary style being English.

One of the earliest settlers were the Amish in the Arthur area. They have survived for years with their simple lifestyle, heavy dependence on their barns for their livelihood, and strong religious beliefs.

The farms in the region had numerous barns built, but as farming methods changed and livestock began being raised in larger feedlots the barns became obsolete. As alternative uses of barns fail or the barns become too expensive to maintain, fewer remain each year. The great rural icons of this district are fading fast. We need to enjoy all the great barns of Illinois while we can.

—Harry Porter

District 5, Union County

This beautiful red barn with the dusting of snow on the ground seems to say welcome home. One can almost smell the bacon sizzling in the pan as you pull in the lane leading to the farmhouse. This barn is located south of Jonesboro.

Photograph and information submitted by Kathy Brey.

District 1, Henderson County

This bank barn, located east of Lomax, was rapidly deteriorating when Tim and Julie Rider acquired it in 1996. Trees were cut away from the barn, structural stabilizing was done, and a new shingle roof was put on. The original design of the barn was for workhorses, but it is now used as a home for a horse and donkey.

Photograph and information submitted by Tim and Julie Rider.

District 5, Lawrence County

Bill and Carolyn Inyart own this barn that was built in 1949 three miles southwest of Sumner. It was built for use by both horses and cattle. The barn was remodeled in 2000 and is now used as a garage and workshop.

Photograph and information submitted by Mark Steber.

District 5, Union County

This once stately barn depicts a way of life that is fading away. It may have weathered many a storm, but it still seems to be hanging on to days gone by. This barn, located between Ware and Jonesboro, was torn down shortly after the picture was taken. The owner thought enough of it to keep a fresh coat of paint on it right up to the end.

Photograph and information submitted by Kathy Brey.

District 5, Richland County

This horse barn, located at the east edge of Olney, was built in 1936. It has three horse stalls and a tack room. It is part of the Albert Nix property and is owned by Bill and Cynthia Shelton.

Photograph and information submitted by Mark Steber.

District 3, Hancock County

Daniel Grant Luce built this barn in Sonora Township in 1868 for both horses and cattle. This barn is forty feet by fifty feet and has twenty-two-inch-thick walls and took nine years to build. The bank barn is made of grout, a mixture of lime, water, and gravel, and has sandstone cornerstones. The accompanying farm home overlooks the Mississippi River and is also made of grout.

Photograph and information submitted by Donald Bumphrey.

District 5, Richland County

Ed and Vandy Shake are the owners of this barn located five miles south of Claremont. The barn was built circa 1940 as a general-use barn for both horses and cattle. It was restored in 1998.

Photograph and information submitted by Mark Steber.

District 5, Richland County

This dairy barn was built in 1908 and is currently owned by Jim and Mary Steber. In 1948 the barn was moved a half-mile northwest of its original location. Another larger barn and a two-story farmhouse were also moved due to the construction of Highway 50. This barn is located one mile east of Olney.

Photograph and information submitted by Mark Steber.

District 3, Hancock County

Henry Hackemack built this barn in 1890 of wooden pegs and hewn beams. When William Wollbrink purchased the farm in 1908 he added a cowshed on the north side. This three-story bank barn, located in Walker Township, has been in the Wollbrink family for four generations.

Photograph and information submitted by Donald Bumphrey.

District 5, Perry County

As you enter the main gate at the DuQuoin State Fair Grounds, you are greeted by the horse barns, which play a huge role in the success of the fair over the years. These barns have been home to some of the world's best-known trotting horses over many decades. The famous Hamiltonian Trotting Race was held here for years.

Photographs and information submitted by Frank Dry.

District 5, Pulaski County

As indicated by the date carved in the stone step, this barn was built in 1923. It is located near Villa Ridge. Orlan McBride who was a commercial fruit and vegetable grower built this barn. The barn is now owned by Al and Wanda Atherton. Its most recent use was for cattle and hogs.

Photograph and information submitted by Lois Herr.

District 5, Randolph County

This barn is over one hundred years old. The hayloft floor was made of fence rails and cornstalks, which still remain today. Nineteen years ago red siding and white trim were added. The barn is located two and one-half miles east of Ellis Grove.

Photograph and information submitted by Lloyd Schoenbeck.

District 5, White County

This old barn with an American flag painted on its front is over one hundred years old. The barn has always been in the Lamp family. Norman Lamp taught Ag and FAA at the local high school for eighteen years and retired in 1989. He passed away in July 2002. This farm, located four miles west of Carmi, is a Centennial Farm. Photograph and information submitted by Mary Lou Lamp.

District 1, Woodford County

The tall doors identify this barn as a horse barn. Arthur Engel and his father, Joseph Engel, built this barn in 1919 on a Sesquicentennial Farm that has been in the same family for 171 years. This barn is located just south of Metamora.

Photograph and information submitted by William and Sandra Christ.

District 1, Lee County

This hand-hewn timber peg barn was built on the homestead of John Hetler. It is the only barn around to have a windmill built into it. The barn is still standing four miles northeast of Dixon and is in good repair.

Photograph and information submitted by Russell and Anna Johnson.

District 1, Carroll County

This is a nineteenth century barn with one roof side longer than the other, a saltbox style. The barn, located west of Chadwick, is well constructed and periodically maintained. It has survived many years of use. The construction is mortise and tenon, with the beams joined together by oak pegs.

Photograph and information submitted by Dave Queckboerner and Harold and Kay Andrese.

District 1, Lee County

This photograph depicts an "Amish Style" barn raising near Franklin Grove. The difference is that the people shown here are not Amish but local volunteers.

Photograph by Robin Lahman, information and photograph submitted by Lois Herr.

District 5, Johnson County

Little information was provided in regard to this barn located near Goreville. This picture was taken in 1976 so it may not still be standing. Because of its unusual shape and design we would be remiss not to include it with the other great barns of Illinois. Photograph submitted by Bill Clark.

District 1, Lee County

This sturdy barn was built in 1900 and used for dairy cows and later for beef cattle, horses, and hogs. The barn has four grain storage bins and at one time had an inside silo. The frame was built using tongue and groove with peg construction. While this barn still has its original wood shingle roof, it has been covered with asphalt shingles. Sitting on top of the roof is a cupola with a cow weathervane. The Bollman children and grandchildren have enjoyed the barn over the years. They used the hayfork trip rope as a swing and built hay forts in the haymow. This barn is located on Harmon Road, near Harmon.

Photograph and information submitted by Anna Stouffer and Gladys Bollman.

District 1, Stephenson County

This large horse and dairy barn was built in 1862 on acres homesteaded in 1846. The stone in the center part was hand quarried from the farm about a mile away in the summer of 1861 by one man and a young boy. Stonemasons came to the farm the following year and laid the stones. The stone part of the barn is estimated to be thirty feet by sixty feet. The height to the eaves is thirty-two feet. The red wood section to the right is a grainary, the red section to the left is a cattle shed, and the white building was used as a milk-cooling house. The barn is located near a railroad track and hobos would often sleep in the barn during the Great Depression. Cows were milked in the barn until the 1960s and then hogs were feed in the barn. William Lamm is the current owner. His son, Gerald Lamm, lives on the farm located four miles northwest of German Valley.

Photograph submitted by Ken Kaishan and information by Gerald Lamm.

District 3, Sangamon County

This barn was constructed for James Alphens Galloway and his wife, Allie (Jones) Galloway. When the barn was completed they invited friends, relatives, and neighbors to a barn dance in the hayloft. The Galloway's three grandchildren and spouses, and two great-grandchildren still reside on the Galloway Centennial Farm located on Galloway Road near Pawnee. Photograph and information submitted by Marie Galloway Marvin.

District 1, Winnebago County

This barn, located on Montague Road near Winnebago, was built between 1910 and 1914. It was restored and the roof replaced in 1998. It is no longer used for livestock but makes a great backdrop for the girls' picture.
Photograph and information submitted by Bob and Susan Mexico.

District 5, Jackson County

This dairy barn with its companion silo was built circa 1930. The current owner is Gordon Parrish. Little information was provided to us for this three-bay barn located on New Era Road near Carbondale.
Photograph submitted by Lois Herr.

District 1, Stark County

The original portion of this big red barn, located seven miles northeast of Toulon, was built in the 1860s. An addition was built circa 1920. The original foundation was built of stone. The addition, erected at a barn raising, rests on a concrete foundation. The same beam and wooden peg construction was used in both parts. The barn measures thirty-six by sixty feet. The completed barn had stalls for nineteen horses, which included one for a dress-up horse used for courting. The barn is known as the Jackson Barn.

Photograph submitted by Karen E. Andrew.

District 1, Bureau County

Little history is known about this brick barn other than it was used as a dairy barn at one time. The barn, located south of Walnut, is well maintained with a new green metal roof.

Photograph and information submitted by Doug Anderson.

District 1, Lee County

This barn is on the Guy Schoenholy farm located on German Road northeast of Paw Paw. No other information was made available.

Photograph and information submitted by Elizabeth E. Merriman.

District 3, Sangamon County

Samuel Newton Galloway built this barn prior to 1878. He and his wife, Dicey Elizabeth (Henkle), purchased the farm in 1854. The farm, located on Galloway Road near Pawnee, is known as the Galloway Centennial Farm.

Photograph and information submitted by Marie Galloway Marvin.

District 1, Lee County

This barn, located on Franklin Road near Amboy, was recently painted red with white trim after being gray most of its life.

Photograph and information submitted by Mike and Joyce O'Brien.

District 1, Ogle County

This barn is unusual in that it is built of limestone. The doors rafters and support beams are wood. The exact location or the owner of this barn was not made available to us, but we felt it should be included because of its uniqueness.

Photograph submitted by Lois Herr.

District 1, Ogle County

This picture is of a round-roofed cattle barn with its attached milk-cooling house, companion silo, shed behind, and corncrib to the side. All the buildings are painted red with contrasting gray shingled roofs. The round hay bales and corn dump just add to the picture. This barn is located on Kishwaukee Road on the east side of the Rock River near Byron.

Photograph and information submitted by Joan Eshleman Velde.

District 1, Winnebago County

This barn was built in 1918 and is 130 feet long and 30 feet wide. The barn, located on Edwardsville Road near Winnebago, is now used for storage of machinery and as a shop.

Photograph and information submitted by Steven and Diana Jones.

District 1, Woodford County

This old horse barn, located near Metamora, was built circa 1930. It had a concrete floor, ten-foot-high glazed-tile block walls, shake shingle roof, and a very large hayloft. The barn was destroyed in 1997 by high winds.

Photograph and information submitted by Torry Gannon.

District 5, St. Clair County

This barn, located at Lenzburg, is a house and home to Jenny Vasquez. The A-frame is the residence and the rest of the barn is her father's shed. The picture below was taken one morning after a winter snow in 2002.

Photograph and information submitted by Jenny Vasquez.

District 1, Knox County

Jacob Wolf built this barn in 1860 and it has remained in the same family for five generations. The supporting pillars for this barn are made of one-by-eight-foot oak timbers. The timbers were hewn and put together with mortise and peg. The rafters supporting the roof are roughly hand-hewn, many with the bark still on them. The haymow has four overhead bins for grain storage. Jacob Wolf is Lance Setterdal's great-great-great-grandfather. This barn, now owned by Maridelle (Wolf) Wallace, is located in Haw Creek Township two miles north of the historic Wolf Covered Bridge near North Henderson.

Photograph and information submitted by Lance D. Setterdahl.

District 1, Winnebgo County

It is rare to find two yellow-painted barns with green roofs and of different designs still being used and in good repair. These barns are located at the intersection of Stillman Valley Road and Kiskwaukee Road near Winnebago.

Photograph submitted by Joan Eshleman Velde.

District 5, Pulaski County

Delbert L. Russell built this barn near Kanark in 1939. His brother Kenneth built a sister barn in 1940 one-half mile away with only slight changes in the inside design. The trusses for the curved roof were built first with one-by-four-inch green oak boards and lapped four boards thick. The foundation was made from concrete and river gravel. The bottom floor walls are five feet tall solid poured concrete with cutouts for the doors. The barn was built with a ten-foot hallway running the length of the barn with stalls on both sides and an inside granary. Both this barn and its sister barn are in excellent repair.

Photograph submitted by Ken Kaishan and information by Charles Douglas.

District 1, Henderson County

This barn is known as the Centennial Barn because it was built in 1876 north of Raritan by the King family. This barn initially housed a very active and large purebred draft horse operation. In 1976, during the bicentennial of the United States, Gene and Marie Missavage renovated the barn to the condition of this 1985 picture. Now owned by Jeff and Vicki Keever, the barn still hosts its greeting sign Cead Meila Failte *which proclaims in Gothic "One-hundred thousand Welcomes." The Keevers presently use the barn to house sheep.*

Photograph and information submitted by Kent Slater. Bill Clark and Lee Brokaw also submitted a photograph.

District 1, Ogle County

This barn was built in 1930 on homesteaded acres. The Waterbury ancestors came west and paid $1.25 an acre for the land, which stayed in the family until 1966, when heir Clarence L. Waterbury sold it to Howard W. and Mary Elizabeth (Harres) Garrison, the present owners. This barn is located on Fairmont Road near Polo.

Photograph and information submitted by Mr. and Mrs. Howard W. Garrison.

District 1, Jo Daviess County

This large red barn with its triple cupolas is used as a dairy barn. Little information was made available to us about this barn located near Willow.

Photograph provided by Bill Clark.

District 2, McLean County

This barn, located near Arrowsmith, has a hex sign on the front. This is a sign of friendship not a symbol to get rid of evil spirits.

Photograph submitted by Bill Clark.

District 1, Lee County

These two red barns with the two companion silos are part of a large cattle-feeding operation. Little information was provided on these barns located near Dixon.

Photograph submitted by Bill Clark.

District 1, Stephenson County

Little information was made available to us in regard to this barn of unusual shape with its companion silo. It was used for a cattle-feeding operation. This barn is located near German Valley.

Photograph submitted by Bill Clark.

District 1, Ogle County

"I cherish my childhood days as memories of mother hand-milking family cows early morning and evening. I remember hooking horses to the haymow rope to pull loose hay up into the haymow by fork. I learned the blessing of work by helping father clean calf pens on Saturday fork by fork into our Great Western manure spreader and taking the spreader load to the field by a team of horses." This barn is located on Lindenwood Road near Kings.

Photograph and information submitted by Norman and John Swanson.

District 1, Lee County

These barns are located two miles south of Sublette. There are two barn pictures submitted. Cornelius Vaessen, grandfather of Lois Vaessen, settled in the Sublette area and built these two barns. One has been re-sided, the other has its original weatherboard siding still exposed to the elements. The barn with the white doors is used to store large round bales of hay. The other barn is used for machinery storage. Photograph and information submitted by Louise Vaessen.

District 1, Ogle County

This red barn with white trim was built in 1877. This date is clearly visible on the front of the barn. The barn was built by Prescott Talbot and owned by the Talbot family until 2001 when it was acquired by the Carmichael family of Maplehurst Farms. In 1995, the barn was struck by high winds, which twisted the frame. The damage was repaired and the barn remains in good condition. The barn is located on Moore Road north of Rochelle.

Photograph and information submitted by Roselyn Nesbit.

District 1, Whiteside County

This old barn with its companion silo is no longer standing but hopefully is not forgotten. It was located on Old Route 2 near Erie.

Photograph and information submitted by Joan Eshleman Velde.

District 1, Ogle County

This barn seems small in comparison to some, but with its stone foundation walls and its weathered red vertical board siding touched off by its bright red door is still a nice scene on the landscape. The age of the barn is not known but the date on a nearby silo is 1906. This barn is located on the River Road east of the Rock River between Byron and Oregon.

Photograph and information submitted by Joan Eshleman Velde.

District 1, Lee County

This barn, located on Lost Nation Road east of Dixon, sits on gently rolling land with a panoramic view of the surrounding countryside. The Hoban family has owned this farm for one hundred years.

Photograph and information submitted by Roselyn Nesbit.

District 1, Winnebago County

In 1925, two barns were put together to form one barn. The original barn is on the left and was built in the 1890s. The barn on the right was disassembled, carried by wagon and team of horses twelve miles, and reassembled. The barn now has a tin roof and metal siding. The barn has been gutted and a large fourteen-by-twenty-foot door was cut in to the original barn. It is now used as a machine shop. The barn is located on Montague Road north of Winnebego. Photograph and information submitted by Steve, Leland, and Diana Jones.

District 1, Knox County

This large black-and-white scene shows a barn and a windmill located northeast of Galesburg. No other information available.

Photograph and information submitted by Bob and Anna Steinman.

District 4, Iroquois County

Still in use today near Stockland, this faded red barn is over one hundred years old. It was constructed using wooden pegs. Photograph and information submitted by Bev Hinkle.

District 1, Winnebago County

This large red barn with its companion concrete silo, wood-shingled roof, and metal air vent has become adorned with modern artwork. This barn has several outbuildings around it, so at one time it must have been part of a large farmstead. This barn is located near the intersection of Route 2 and Business 20 Bypass at Rockford.
Photograph submitted by Joan Eshleman Velde.

District 1, Ogle County

This barn is a good example of a stone foundation barn. This barn has an attached loafing shed and a companion silo. The barn sports a metal roof and metal air vents. It appears that it is having a color change to gray. This barn is located on Stillman Valley Road on the east side of the Rock River.

Photograph submitted by Joan Eshleman Velde.

District 1, Winnebago County

The Kent Benson barn was built in 1889 alongside a creek on an early rural farmstead. Jerome Doeden was the previous owner and owned it for many years. The barn is currently being used for hay storage. This is one of the few barns left with a wooden hay track. This barn is located on Skare Road near Chana.

Photograph and information submitted by Norma Swanson.

District 1, Bureau County

This barn, located south of Walnut, has post and pegged-beam construction. The lower level is constructed of tile block. It is a bank barn that was constructed in 1923 from a Sears and Roebuck kit by Syd and Mabel Fordham. Owner Steve Gerdes has restored the barn to its present condition. Photograph and information submitted by Steve Gerdes.

District 1, Bureau County

This barn has post and pegged-beam construction. The lower level is glazed tile. Larry and Ann Gerdes, the owners, have restored the barn to its original condition, even leaving the dairy facility intact. This barn is located northeast of Walnut. Photograph and information submitted by Larry Gerdes.

District 1, Ogle County

This large barn was once the central part of a thriving active farmstead, but obsolescence has taken its toll. This barn is located on the River Road on the east side of the Rock River between Oregon and Byron. Photograph and information submitted by Joan Eshleman Velde.

District 1, Whiteside County

This picture stands out because the picture is taken through the limbs on an old tree. Both are about the same age it appears. This barn is the center point as it sets on a hill overlooking a horse pasture, surrounded by trees. This barn is located on Old Route 2 near Erie.

Photograph and information submitted by Joan Eshleman Velde.

District 1, Ogle County

This big barn is titled Pulpit Rock Farms Barn No. 2. It is located on West Edgewood Road near Dixon. It is believed that this large red barn was built around the turn of the twentieth century.

Photograph and information submitted by Randall and Peggy Renne.

District 1, Henderson County

The walls of Clayton Stambaugh's barn, located near Lomax, are constructed of native rock. The barn has a hip roof. The actual date the barn was built is unknown, but it is believed to be in the 1860 time period. It is used today for grain storage and livestock. Photograph and information submitted by Kent Slater.

District 1, Ogle County

The red barn in the foreground was built in the 1890s. The barn in the background was built in the 1960s. These barns are located south of Polo. Photograph and information submitted by Curt and Pat Erickson.

District 1, Woodford County

This large round-roofed barn and companion corncrib attract the attention of travelers on Route 150 between Goodfield and Carlock. Photograph submitted by Anne and Bob Steinman.

District 1, Stephenson County

This old barn, located on Pearl City Road near Pearl City, was converted to an antique shop and party barn. Until his death, Lee Hanson had a wood stove, antiques, jukebox, and parties in this barn. Photograph and information submitted by Sharon Hull.

District 1, Lee County

The cold winter sky and the snow-covered ground with the frosted trees help accent this freshly painted red barn with its white trim—surely a "Kodak Moment" in any barn lovers mind. This barn is located on Franklin Road near Amboy.

Photograph and information submitted by Mike and Joyce O'Brien.

District 1, Mercer County

This faded white barn, located on Great River Road at the north edge of New Boston, has a full basement.

Photograph submitted by Paul Phelan.

District 2, Iroquois County

This barn, located near Gillman, was built in 1926. It is located on Green Acres Farm. The barn and farm have special meaning to the owners Jim and Carol Spielman. While they both came from farming families, they are the first to return to the farm.

Photograph and information submitted by Carol Spielman.

District 2, Kankakee County

This barn, located north of Kankakee, has a dormer door above the milking area of the barn and a milk-cooling house attached. The red paint and the green roof set it off. The concrete silo with its metal support straps adds to the picture.

Photograph submitted by Paul Phelan.

District 2, La Salle County

This barn, built in 1912, is known as the Maier Farm Yellow Barn.
The exact location of this barn is not known.
Photograph and information submitted by Paul Phelan.

District 1, Carroll County

The overall effect of this picture is an excellent example of how the barns of Illinois add beauty and character to the landscape, even though this barn is not unusual. This barn is located between Lanark and Milledgville.
Photograph and information submitted by Jeff Slamp.

District 1, Ogle County

This well-kept barn was built in 1937 as a horse barn. Later it was converted to a dairy barn. It continues to be used for six Warner children's various 4-H animals. The haymow is still used to store hay and straw. A metal roof has recently been added to help preserve the barn which is located on Stillman Valley Road near Stillman Valley.

Photograph and information submitted by Joy and Mark Warner and Norma Swanson.

District 1, Winnebago County

Don and Joan O'Rourke converted this gable-roofed barn into a spacious residence. Now it is a charming bed and breakfast called the Barn of Rockford owned and operated by Ken and Karen Sharp. It is directly across the road from Rockford's Midway Village and Museum Center on Guilford Road.

Photograph and information submitted by Gail S. Ravitts.

District 3, Mason County

This old barn, located near Bath, has a ramp entranceway to the main floor and a full basement. The roof supports three metal vents. This picture was taken in 1976.

Photograph submitted by Bill Clark.

District 3, Adams County

This barn, built in the 1950s, replaced the previous barn which had been destroyed by fire. The barn was formerly used to house dairy cattle and swine. Currently it is only used to store hay. In an average summer over five hundred bales of hay are stacked in its haymow. The barn is located near Camp Point.

The photograph submitted by Henry Bockholder, information by Michael Bockholder.

District 3, Hancock County

Lance and Layne Roskamp, the current owners of this barn near Sutter, purchased it from Charlie Hutchinson's estate in 1979. The barn has two grain bins and a large haymow.

Photograph and information submitted by Jami Roskamp.

District 3, McDonough County

No information was provided for this barn but the winter landscape makes for a beautiful scene. This barn is located near Bushnell.

Photograph submitted by Bill Clark.

District 1, Ogle County

This gray barn with its many glass windows was constructed entirely of cypress in the 1930s. Architect Phillip Drummond designed this barn located on Edgewood Road near Dixon. It was used as a show barn and stable.

Photograph and information submitted by Randall and Peggy Renne.

District 2, Livingston County

This gray stone block barn with its curved windows looks like a church. This frequently photographed barn is located at Kangley Road and Route 17 near Streator.

Photograph and information submitted by Paul Phelan.

District 1, Whiteside County

As the photographer rounded a curve she spotted this barn with some roof damage. She hopes the roof has been repaired, as this is a nice strong barn, located a few miles west of Prophetstown.

Photograph and information submitted by Roselyn Nesbit.

District 5, Massac County

Donald Russell purchased this farm in 1961. The barn's actual age is not known. It was built as a horse and dairy general-purpose barn. The barn still has intact one large horse stall and two smaller ones. The barn is forty-six by seventy feet including the lean-to. The barn is constructed using mortise and tenon construction with wooden pegs. This barn is located six miles northeast of Metropolis.

Photograph submitted by Ken Kaishan and information by Donald Russell.

INDEX

District 5, Massac County

Donald Russell purchased this farm in 1961. The barn's actual age is not known. It was built as a horse and dairy general-purpose barn. The barn still has intact one large horse stall and two smaller ones. The barn is forty-six by seventy feet including the lean-to. The barn is constructed using mortise and tenon construction with wooden pegs. This barn is located six miles northeast of Metropolis.

Photograph submitted by Ken Kaishan and information by Donald Russell.

INDEX

DISTRICT 1

DISTRICT 2

DISTRICT 3

ABOUT THE AUTHOR

Photograph by Gene Schniele

Harry Porter was born in 1935, the seventh and youngest child of W. G. and Flossie Lamb Porter, and was raised on a farm south of Bement, Illinois. He was attending Eastern Illinois University when his father died. He returned to the farm and helped his widowed mother. In 1962, he became a Country Companies Insurance agent, retiring in 1999 after thirty-eight years of service. He married Carol J. Heckman in 1967 and they are the proud parents of two sons.

He has been a member of the First Presbyterian Church for over fifty years. He sings in the church choir and is a past elder and deacon.

He is active in the Bement Masonic Lodge 365 A.F.&A.M., serving as master in 1969, and currently serves as senior steward. He is a member of the Bement Royal Arch Chapter No. 65, and currently serves as sentinel.

He was a member of the Illinois National Guard from 1957 to 1964 and is a member of the Albert Parker American Legion Post No. 620. He is co-chair of the Bement Area Veterans' Memorial Committee that received the Points of Life Award.

Porter is currently serving as president of the Bement Public District Library Board and is a member of the Friends of the Library. He is a life member of the Bement Parent Teachers Association, a participant in the Dear Reader Program, a volunteer for Bement School's Learning Disabilities Program, and is an active certified mentor.

He has been a member of the Piatt County Farm Bureau for over forty-five years. He has also served on the Bement Village Board, the Bement Community Council, the Eastern Illinois Tourism Council, and has been a Boy Scout leader. He was inducted into the Bement High School Alumni Hall of Fame in 1999. He is currently the treasurer for the Bement Area Chamber of Commerce, a Board member of the Piatt County Mental Health Center, and is serving on the Advisory Committee for the 2005 Bement Sesquicentennial.

Porter's hobbies include high school football, barn photography, traveling, reading, drawing rural scenes and barns (example below), painting ceramic pieces, lecturing on barns, and gardening. He is active in the Master Gardner Program in Piatt County. He served as spokesperson for the Piatt County Barn Tour (1998) and served on the advisory committee for the 2002 Piatt County Barn Tour.